THE REMOTE SENSING SOURCEBOOK

A guide to remote sensing products, services, facilities, publications and other materials

THE REMOTE SENSING SOURCEBOOK

A guide to remote sensing products, services, facilities, publications and other materials

D. J. Carter

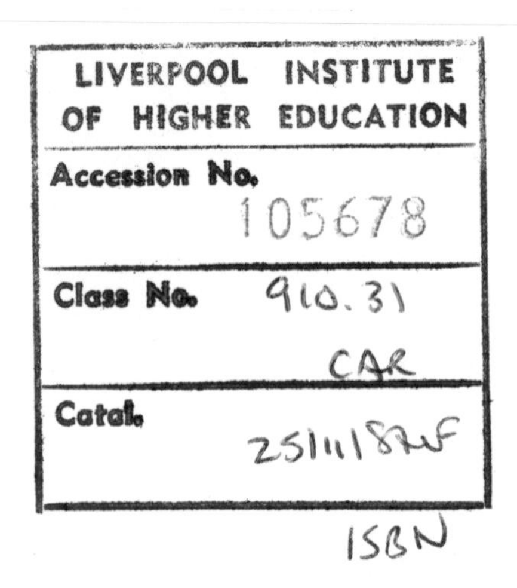
LIVERPOOL INSTITUTE OF HIGHER EDUCATION
Accession No. 105678
Class No. 910.31 CAR
Catal. 25/11/87
ISBN

First published in 1986 by
McCarta Ltd, 122 King's Cross Road,
London WC1X 9DS and Kogan Page Ltd,
120 Pentonville Road, London N1 9JN

All rights reserved. No part of this book may be reproduced or transmitted in any form or by any means electronic or mechanical, including photocopying, recording or by any information storage and retrieval system, without permission in writing from the publisher.

British Library Cataloguing in Publication Data

Carter, D. J.
The remote sensing sourcebook: a guide to remote sensing products, services, facilities, publications and other materials.
1. Remote sensing systems
I. Title
621.36'78 G70.4

ISBN 1-85091-034-0 - Kogan Page

ISBN 0-906318-15-7 - McCarta Ltd

Printed and bound in Great Britain
by Anchor Brendon Ltd, Tiptree, Essex

'You, Sir...will easily conceive with what pleasure a philosopher, furnished with wings, and hovering in the sky, would see the earth, and all its inhabitants, rolling beneath him, and presenting to him successively, by its diurnal motion, all the countries within the same parallel. How it must amuse the pendant spectator to see the moving scene of land and ocean, cities and deserts!

To survey with equal security the marts of trade,... mountains infested by barbarians and fruitful regions gladdened by plenty. How easily shall we then examine the face of nature from one extremity of the earth to the other.'

Dr Samuel Johnson, History of Rasselas, 1759

CONTENTS

FOREWORD

It is the purpose of this book to provide a clarification of the organisational structure of remote sensing activity in the UK and to give a basic outline of the products, services, facilities and publications that are available. The need of teachers, lecturers, advisers and other members of the educational community for a basic orientation in this field has been the prime concern in compiling and editing the contents. It is not intended as an encyclopaedic directory for specialists and professionals in remote sensing, although it is hoped that some of them may find this a useful summary. Image processing and other quantitative analytical methods and equipment are therefore given only relatively brief treatment.

Although the emphasis is on products, publications, services and facilities in the UK, it would be unrealistic, and arbitrary, to confine this guide to a British context alone. Remote sensing, particularly from space, has an international dimension, and progress in the UK has been wholly dependent on access to data provided via American and European networks and agencies. Imagery archives, publications, etc. that are readily available from these sources are therefore acknowledged. For the same reason, texts, audio-visual resources and maps published overseas are also included on a selective basis.

Space has prevented the inclusion of review comments on all published texts and other resources, although an indication is given of those products which have been widely adopted or which offer flexibility in their use as teaching and learning materials. Detailed reviews of many of the books and some of the audio-visual resources have appeared in journals and newsletters that are listed later in the Technical Literature section.

Copyright on all products remains with the original publishers. Although space imagery from American sources is in the public domain (i.e. it is not subject to copyright), this does not apply to imagery that has been processed in any way by other agencies. Anyone wishing to reproduce imagery for commercial purposes must first determine any copyright restrictions.

Details are correct at the time of going to press; any inaccuracies are the responsibility of the author, who would be pleased to receive details of omissions and errors that may be apparent to readers. The Addenda lists information that became available after the final proof copy of the text had been completed.

Acknowledgements

The chapter on satellite image maps was compiled jointly with McCarta Ltd. Dr Ross Reynolds, Department of Meteorology, Reading University contributed most of the section on satellite meteorological data.

Valuable comments on the initial draft were received from Miss P. A. Vass and Dr R. K. Bullard, of the National Remote Sensing

Centre and Dr P. Collier, Department of Geography, Portsmouth Polytechnic. The clarifications and corrections they provided are gratefully acknowledged. Mr L. Gold also contributed comments on the teaching potential of Meteosat imagery.

Typing of the manuscript was undertaken by Mrs Margaret Bristowe, Mrs Christine Carter and Mrs Sandra Winterbotham. Christine Carter also compiled the Index and List of Abbreviations, and helped in many other ways in accelerating progress at all stages in the compilation of this book.

A special debt of gratitude is due to Mr J Henderson McCartney, Managing Director of McCarta Ltd. Without his initiative, and his vigorous interest, encouragement and advice this book would not have been written. The detailed guidance and help provided by Mr Kevin White, of Kogan Page, is also gratefully acknowledged.

D. J. Carter

Havant, Hampshire
January 1986

To avoid repetition, the addresses of organisations having several entries are given only once. The relevant page is underlined in the Index.

Part I

An Overview of Remote Sensing Activity

PART 1.1: AN INTRODUCTION TO REMOTE SENSING

This brief summary of the main concepts and methods of remote sensing is designed to give basic orientation for readers without pre-existing familiarity with this field. It will also serve to introduce the main remote sensing programmes and sensor types that are used without further explanation in the text. The list of abbreviations given later should be consulted for additional clarification.

A comprehensive and literal definition of remote sensing is the observation and measurement of the attributes of objects, independent of direct physical contact. As this embraces fields of investigation as diverse as radio astronomy and forensic medicine, it has been reduced in scope to provide a manageable context. In the environmental sciences, remote sensing is understood as the acquisition, recording, processing and classification of data obtained through the use of electromagnetic radiation sensors. The latter may be fixed or mobile ground based systems or accommodated in aircraft, helicopters, rockets, spacecraft or orbiting Earth satellites. In the popular imagination, remote sensing is often equated with space-borne measurements of the Earth's atmosphere, and land and water surfaces. However, these techniques have also been very successfully deployed on deep space planetary missions since the early 1970s.

Sensors adapted to certain parts of the electromagnetic spectrum are not the only types available. Useful environmental data may be obtained from the exploitation of acoustic wave and force fields, and from sensors that are not designed to produce any form of imagery from their fields of view. These are, however, excluded from the scope of this guide.

Electromagnetic radiation occurs as a continuum of wavelengths and frequencies (FIG 1). Those most commonly used in operational remote sensing are in the visible and near infrared waveband (0.4-1.1μm); infrared, including thermal infrared, (3-14μm) and microwave (1-500mm). The amount, and characteristics of, radiant energy reflected and emitted from the Earth's varied surface cover depends on the physical and chemical characteristics of specific objects. Sensor systems designed to record this within one or more defined wavebands, are known as passive. Those systems (of which radar is the best known) that use artificial electromagnetic energy sources as the basis for image construction are referred to as active. There are a large number of concepts and terms that describe the energy radiated from any object or set of objects within a unit area. That of radiance is the most important, as it quantifies the amount of energy radiated and recorded by a sensor within a given observational 'window'.

Visible light and near infrared radiation, reflected by objects at the Earth's surface, can be recorded on photographic film, and both vertical and oblique aerial photos constituted the dominant remotely sensed image product up until the mid-1960s. Thereafter, increasingly more versatile and sensitive non

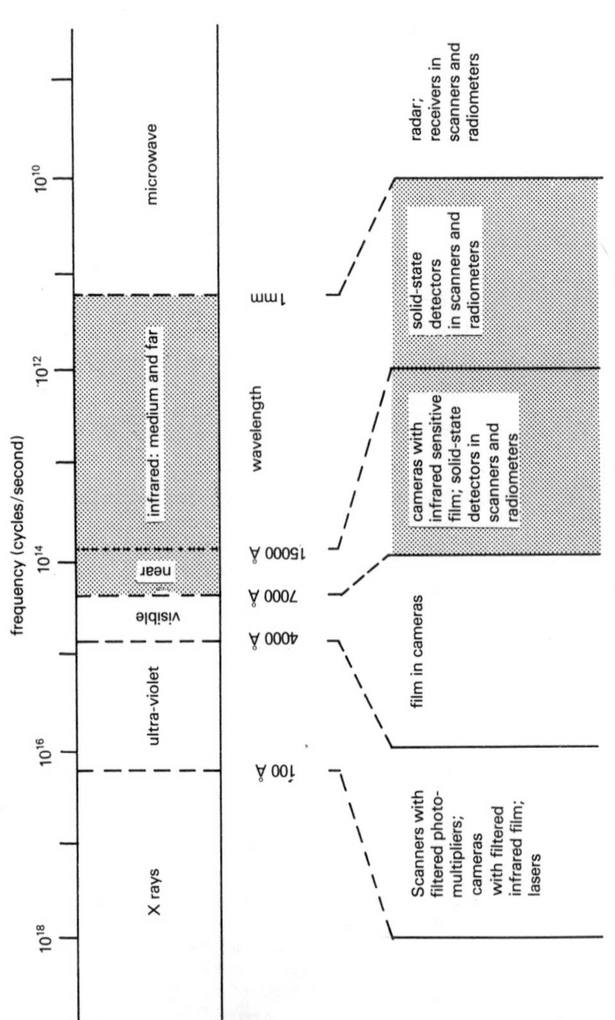

Figure 1. The electromagnetic spectrum and some common sensors that use it (adapted from Holz, 1973).

photographic systems were designed to record both reflected and emitted energy, such as line-scan (FIG 2). This development coincided with the design and launch of automatic satellites, so that the capability of remote sensing increased dramatically. Not suprisingly, the major technological advances throughout the 1970s, into the 1980s, have been concerned with improving the capabilities of space based remote sensing in terms of rates and quality of acquisition, classification, storage and retrieval of information. Computer-assisted techniques of image processing for the purposes of interpretation of data for a very wide range of applications have also developed rapidly.

That does not imply that conventional photographic remote sensing has been relegated to a 'second class' position. It remains a primary technique where it is necessary to resolve the detail of ground conditions, as in most forms of resource and environmental planning. More recently, photographic sensors have been deployed on board Space Shuttle missions, with very promising results. True colour photographs of large areas of the Earth, taken with hand held cameras on manned spaceflights in the late 1960s and early 1970s constitute an important visual record that continues to stimulate public awareness of remote sensing.

Aerial remote sensing has its limitations, however. Not least of these is the logistics of operating an aircraft, involving personnel, expensive equipment and suitable flying and sensing conditions. Not surpisingly, there are few areas in the world for which there is repetitive cover of airborne imagery at comparable scales. Orbiting satellites offer the considerable advantage of acquiring data of a given field of view of the Earth's surface at a known temporal frequency and in a uniform format. It is true that satellites dedicated to imaging in the visible and near infrared spectral regions are dependent on natural illumination and on clear atmospheric conditions, so that the actual frequency of repetitive cover may be considerably less than the nominal interval between 'passes'. The recent, and future, development of active microwave and thermal infrared sensor systems will help to modify this constraint, as both have an all-weather, day and night time capability. Both can be deployed in aircraft as well as on space platforms, and have already proved their value in contexts, such as large area regional surveys of vegetation (radar) and the detection of geologically significant linears (thermal infrared). Thermal infrared remote sensing records the temperature characteristics of objects, and can be used to either estimate or detect temperature differences as a basis for discrimination between surface (and indirectly, subsurface) phenomena. Microwave remote sensing, especially synthetic aperture radar probably has the greatest immediate potential for synoptic mapping of oceanic, ice and other water surfaces. Distinction between surface features on radar imagery is dependent upon the interaction of microwave energy and the geometry of surfaces (relief interval, angularity or curvilinearity of form, particle size of sediments, etc) and is therefore both distinct from - and much more complex than - the recording of reflected and emitted electromagnetic energy at shorter wavelengths.

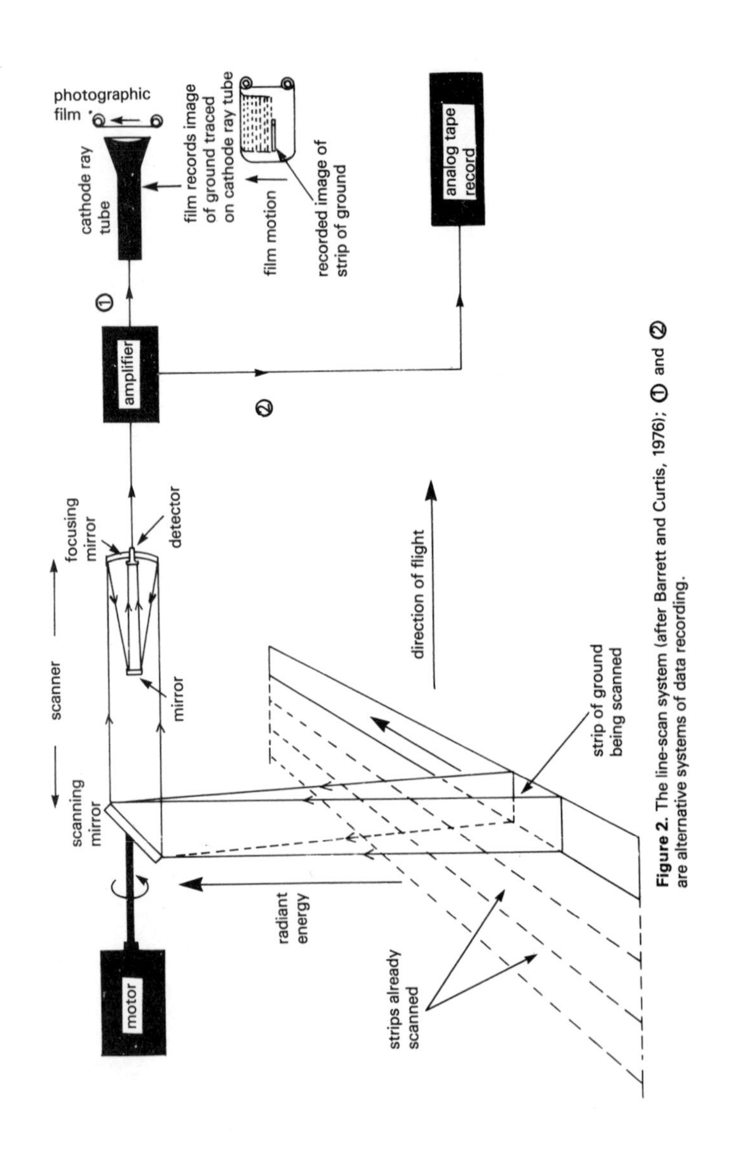

Figure 2. The line-scan system (after Barrett and Curtis, 1976); ① and ② are alternative systems of data recording.

There are other reasons apart from operational advantages that have favoured satellite-based remote sensing. Because these systems are automatic, data is recorded as digital code and transmitted to ground receiving stations. This is written on to computer tapes, in which form it is available for a multiplicity of processing or analytical techniques that can be applied to the 'raw' data. The rapid development of the capacity and capability of computers in the past 15 years has coincided with the availability of satellite data of increasing quantity, variety and complexity. This mutual relationship has promoted innovative development of image processing techniques appropriate to the research and applications interests of the remotely-sensed data user community (eg public agencies, planning units, commercial organisations such as hydrocarbon and mineral prospecting companies, and academic institutions). Although there are no limitations to the use of digital multispectral scanning systems on aircraft, sets of aerial imagery of this type are uncommon by comparison. There is a straightforward reason for this: most aircraft are operated commercially, and remotely-sensed data obtained from air survey is in the copyright ownership of companies or their clients. In contrast, satellite data has been placed in the public domain, ie it is not subject to copyright (at least not in its 'raw' form; second generation processed products may be). United States policy has been one of 'open skies', which has allowed the reception and analysis of data from American satellites free of ownership restrictions or overt political control. There are growing indications that this policy may not persist, and it only applies, of course, to civil remote sensing programmes.

All remotely-sensed images, whether viewed on the visual display unit of a computer or converted into 'hard copy' photographic print form are characterised by three resolutions - spectral, spatial and temporal. Together, they determine the discrimination between, and analytical significance of, the units of data that compose the image. (In the case of radar images, the nature of the interaction between microwave radiation and characteristics of the surface is an important though not the only determinant of resolution). Spectral resolution involves the degrees of contrast in radiance that are characteristic of different materials and objects in defined wavebands of the electromagnetic spectrum. It will vary not only according to sensor design, but with prevailing environmental conditions (eg whether the surface is wet or dry). Spatial resolution is a function of the size of the data collection units characteristic of the sensor and it ultimately determines the detail that can be extracted from imagery. Thus if the nominal spatial resolution is 80m, no object of smaller plan dimensions will resolve. It should be noted that spatial resolution need not be a function of the altitude of the sensor; the popular concept that the higher one goes the less detail one can discern is only of limited relevance. Temporal resolution is simply a function of environmental conditions prevalent at the moment of imaging. A rainstorm or smoke plume, for example, may make an important contribution to the information on an image, and there are also predictable temporal changes that are of vital significance in selecting imagery for interpretation. Seasonal changes not only involve distinctive differences in vegetation cover, but are likely to affect river regimes, snow lines, etc, as well as human activities. A sequence of images for a given location or region, chosen to detect longer term natural or socio-economic changes, must be carefully selected with this factor in mind. Natural illumination will also vary both systematically and randomly and may either enhance or suppress the visibility of features such as relief forms.

It is for this reason, amongst others, that remote sensing cannot be practised without knowledge of field conditions. Both visual (manual) and computerised image analysis and interpretation has to relate to knowlege of what is called in remote-sensing jargon, 'ground truth'. In many contexts, field data can be collected on a sample basis, so that the role of remote sensing is to provide informed identification and classification of areas not directly measured. For physically or politically inaccessible regions, or where field survey costs would be very high, the time and cost-effectiveness of all forms of remote sensing becomes obvious. Not surprisingly, both aerial and space remotely-sensed data have been attractive in a very wide range of environmental and resource surveys in both developed and developing countries, as well as at global scales. Surveys of crop conditions, forest timber, water quality, snow and ice volumes, urban area development and mineral extraction are just a few of the numerous applications that have used a judicious combination of field and remote sensing measurements as a basis for management and planning.

Throughout this book, the names of certain satellite remote sensing programmes are used without background explanation. They are briefly summarised here for the convenience of readers.

A. MANNED SPACE PROGRAMMES

1. Gemini and Apollo

NASA manned spacecraft programmes which included experimental true colour photography of selected areas of the world. The Gemini missions obtained nearly 2,500 photographs, mostly oblique views, between 1965 and 1966. The Apollo flights were mainly dedicated to the objective of lunar landings, but Apollo 6 and 9 (1967) included experiments in automatic colour photography and multiband imagery that proved important in the later development of the Landsat satellite system.

2. Skylab

A space station, operated by NASA between 1973 and 1974, to accomodate a wide range of space-related experiments. These incorporated a multicamera array and a 13-channel multispectral scanner. The former provided imagery at spatial resolutions of between 30 and 80m, which is still of value to researchers and applications specialists. Over 35,000 scenes were obtained from the various camera systems operated at different times.

3. Space Shuttle

A series of four NASA spacecraft that eliminate much of the wastage of hardware at launch and re-entry; it is an essentially re-usable, long lifespan vehicle (FIG 3). The orbiter vehicle has deployed a number of important remote sensing experiments during the 11 missions from 1981 to date (1985), including the Spacelab 'package' designed by the European Space Agency. Missions are normally 5-7 days in duration and obtain remotely-sensed data for research and development purposes. Of special interest has been the space testing of the MOMS and large format metric camera, and the acquisition of shuttle imaging radar data. The latter is a synthetic aperture radar system with a capability of providing calibrated imagery at an approximate 40m spatial resolution. Once the Space Shuttle is routine, it will provide the opportunity for short-term as well as repetitive remote sensing surveillance and monitoring. It also has the function of launching satellites and retrieving others for repair and servicing, although recent missions have encountered problems in this respect.

B. AUTOMATIC (UNMANNED) SATELLITES

1. Landsat

Originally called ERTS (Earth Resources Technology Satellite), the Landsat programme acquired its present acronym in 1975. Landsat-1 was launched in 1972 and was equipped with a RBV (Return Beam Vidicon) television camera system and a four waveband Multispectral Scanning System (MSS). The latter operates in two channels of the visible spectrum and two in the near infrared. It records radiant energy from the field of view by means of on-board detectors that converts this, electronically, on to magnetic tape (it can also be telemetred directly to ground receiving stations) (FIGS 4 and 5).

Landsat 2 (launched 1975) and 3 (launched 1978) were virtually carbon copies of the prototype. Landsats 1, 2 and 3 had an orbit height of 910km, and a capacity to image the entire Earth's surface (below 81°) every 18 days.

Landsat 4 and 5 (1982, 1984) incorporate important improvements. They orbit at 705km and have a global repeat imaging frequency of 16 days, but also carry a seven waveband sensor known as the Thematic Mapper (TM). Nominal spatial resolution of images obtained from six of these channels is 30m, which compares to 80m for the MSS system. From the start of the programme, Landsat data has been made available to the worldwide user community free of access or copyright controls. This, together with the quality and versatility of information made available, has been of great importance in promoting the civilian use of remote sensing from space in the last decade (1975-1985). The programme has so far been developed and administered by public agencies in the United States, but there could be radical impacts on access and cost-effective use of Landsat data, especially for institutions with small budgets, once Landsat becomes a commercial enterprise (see Addenda).

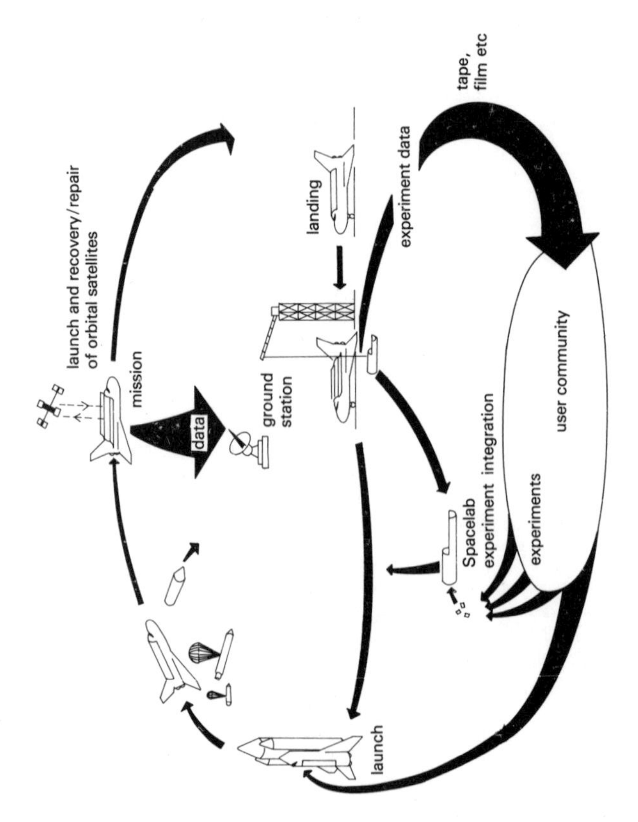

Figure 3. The remote sensing capability of the Space Shuttle/Spacelab system (adapted from Barrett and Curtis, 1983).

2. Heat Capacity Mapping Mission (HCMM) Satellite

Operational between April 1978 and September 1980, this experimental NASA satellite was equipped with a scanning radiometer that obtained data in visible, near-infrared and thermal infrared wavebands. Image resolution varied from 0.6 to about 1km, over a field of view of approximately 700 x 700km. The orbital configuration was designed to maximise opportunity to gain data on surface temperatures. Geological and vegetation mapping, snow and ice melt prediction, and industrial pollution are amongst some of the applications that have made positive use of HCMM data.

3. Seasat

Although operated by NASA as an experimental satellite, the data from this three month mission in 1978 has proved of great interest to earth and environmental scientists. Seasat carried a Synthetic Aperture Radar (SAR) and a radiometer that recorded in visible and thermal infrared bands. It is the SAR imagery that has attracted most research attention. Although, as the name implies, Seasat was primarily designed to obtain information on the world's oceanic surfaces (wave heights, patterns, sea ice, etc) it has provided original data on land cover. The satellite orbited at a height of 800km, covering the Earth's surface between 72° north and south. The SAR sensor gave imagery in 100km wide swaths at a nominal resolution of 25m.

4. Meteorological Satellites

The more important programmes, and the characteristics, of the satellites involved, are described later. Between 1960 and 1977, all meteorological satellite data available in the west were operated by the USA for both research and data delivery purposes. In 1977, ESA's Meteosat (FIG 6) was launched as a contribution to a network of six geostationary satellites intended to provide coverage of the Earth's surface (between 60° N and S) on a continuous basis. Each satellite covers an area of about 70° of longitude and orbits at 36000km above the Equator. In 1978, the Japanese contribution was added, known as the Himawari. The United States satellites in this same programme are called GOES (Geostationary Operational Environmental Satellites). GOES West covers the Pacific Ocean and the western Americas; GOES East the eastern Americas and eastern Atlantic. Polar orbiting meteorological satellites operate between 500 and 1500km in height, each completing an average of 15 100-minute orbits every 24 hours. They image every point on the Earth's surface twice daily, once at night and once during the day (geostationary satellites by contrast obtain an image of the unchanging field of view every 30-40 minutes). The major polar orbiting programmes have been TIROS (1960-1966); ITOS (Improved TIROS Observational Satellite)(1970-1976); TIROS-N (1979-present) and NOAA satellites (two series) (1973-present). These have been, or are, operated by the US National Oceanic and Atmospheric Administration. NIMBUS 1-6 (1964-1981) was operated by NASA as it had a research and development function as well as

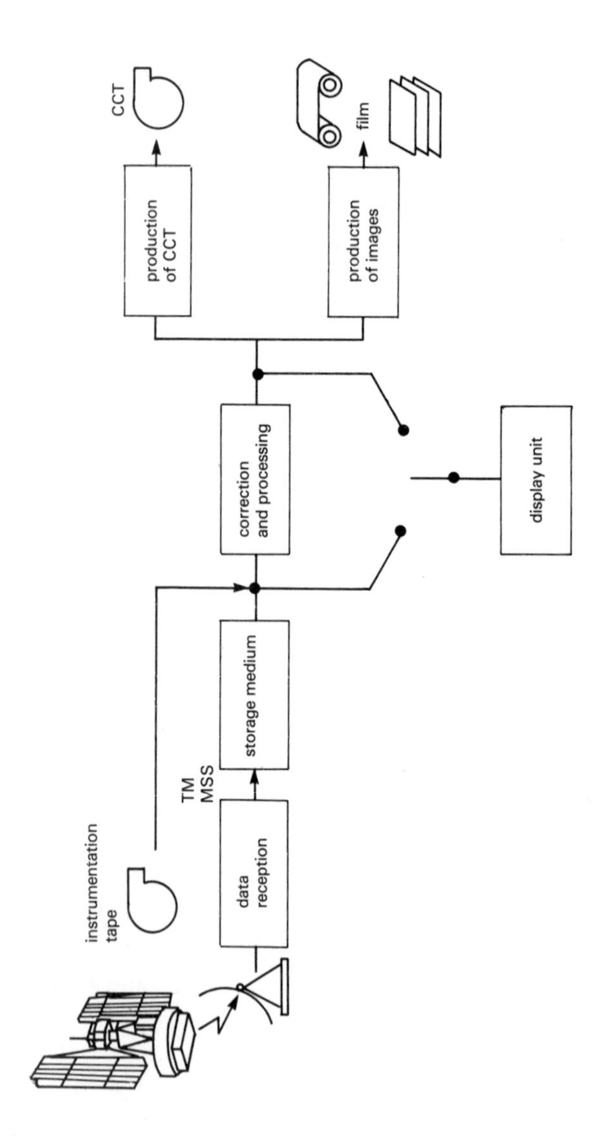

Figure 4. The Landsat system (source: Earthnet, 1978.)

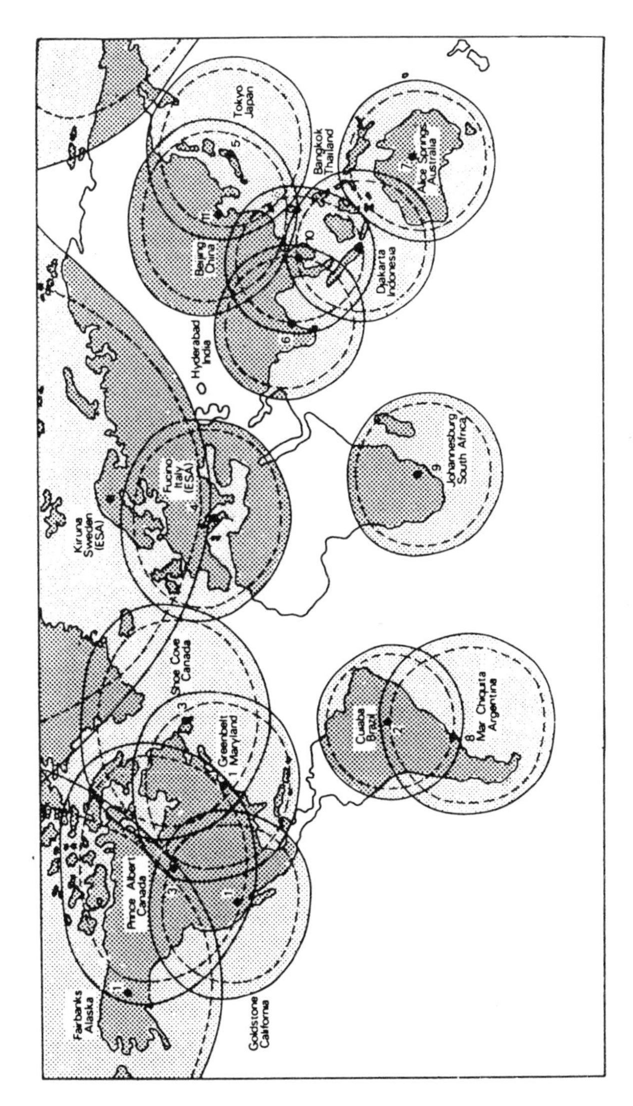

Figure 5. Landsat receiving station coverage: Landsat 1, 2 and 3 ____; Landsat 4 and 5 -----; Receiving station location ● (source: NASA, 1984).

contributing routine meteorological data. NIMBUS-7 (1978-1984) carried a multispectral scanner, the Coastal Zone Colour Scanner (CZCS), dedicated to the monitoring of ocean and coastal waters for organic and mineral suspensates.

5. SPOT

This is the first civilian Earth resources remote sensing programme to be developed independently of the USA and USSR; the first satellite in this series is scheduled for launch in November 1985. It will have a sun-synchronous, near-polar orbit at an altitude of 832km (cf. Landsat), and will achieve a global survey in 26 days. The main sensor is a 'pushbroom' scanner, giving several advantages over conventional multispectral scanning devices (such as improved radiometric and geometric accuracy, and higher spatial resolution). The two scanners in SPOT-I will record in both panchromatic and multispectral modes, the former giving an optimal spatial resolution of 10m, the latter 20-25m. A particular advance of the SPOT system, compared with Landsat, is its potential capability of 'off-nadir' viewing, i.e. the sensor can be directed to + 27° either side of the field of view defined by the vertical axis between the ground and the satellite (FIG 7). This means that the frequency of repetitive cover can be lowered to one to three days (dependent on latitude). Stereoscopic imagery will also be obtainable. The possibilities for monitoring rapidly developing or changing events are considerable. SPOT has been developed primarily by the French Space Centre (CNES), with contributions from Sweden and Belgium. It will supply the global remote sensing data-user community on an open access, commercial basis.

6. USSR Satellites

Several series of both manned spacecraft and automatic satellites have been operated by the USSR since the late 1960s, including the Meteor and Cosmos programmes. The latter has been equipped with multiband photographic, multispectral scanning and microwave sensors, but details are not available in the west. As only a few selected images have been published (eg the atlas referred to on p. 79), and there is no means of access to the imagery archives for western scientists and other users, no direct reference can be made to the Russian contribution to global remote sensing in this book. It is known that the main thrusts of application development have been confined to the national territories of the USSR and COMECON states, with the exception of global meteorological and oceanographic sensing.

7. Military Satellites

It is widely acknowledged that the spatial and spectral resolutions of satellites operated primarily for geostrategic and military reconnaissance purposes are superior to those dedicated to civilian uses. As the imagery obtained is classified there is no means of access, though it is at least comforting to consider that technological advances in the

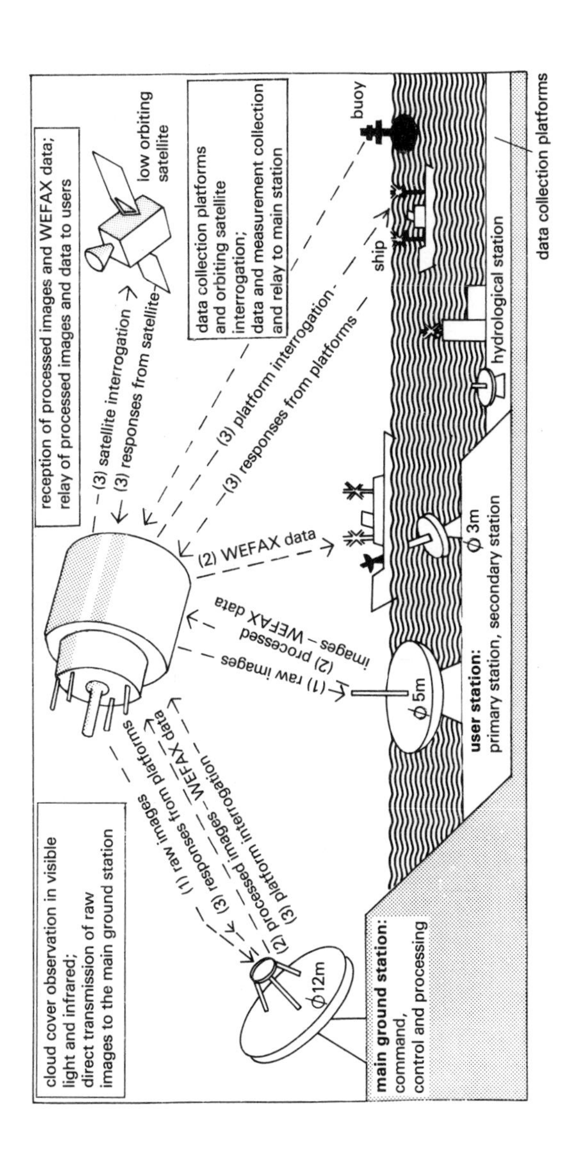

Figure 6. Meteosat data collecting system (source: ESA, 1982).

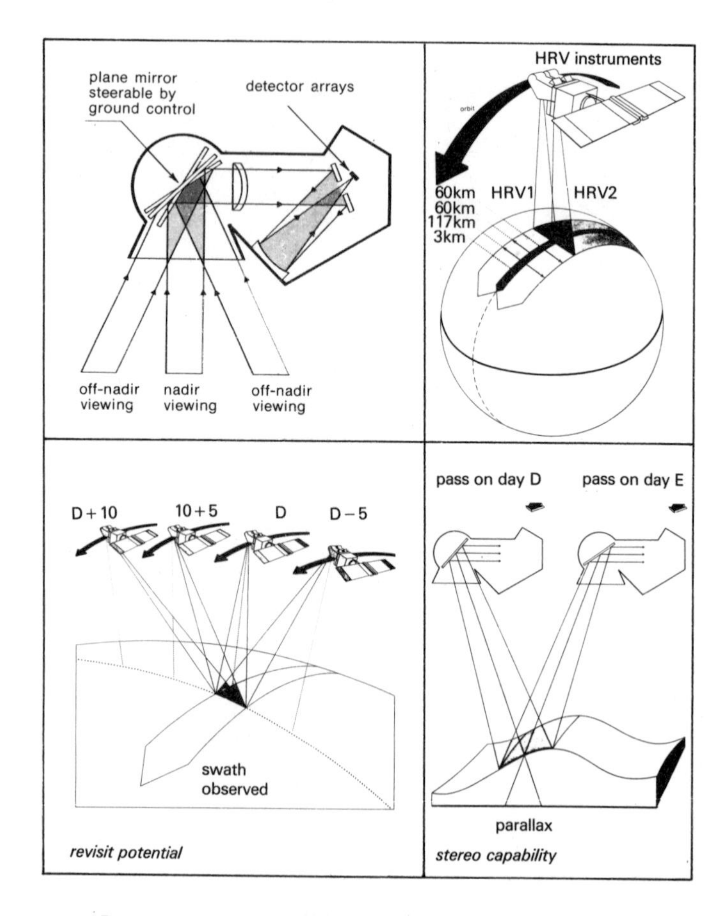

Figure 7. Attributes of the SPOT system (after SPOT-IMAGE and NCRS).

design and operation of military remote sensing systems have contributed directly to the development of civilian systems.

8. Future Civilian Remote Sensing Satellite Programmes

Several satellites are currently in active planning, development and construction. They include the European Space Agency's ERS (Earth Resources Satellite) programme which will include a synthetic aperture radar dedicated to monitoring oceanic conditions. ERS-1 is due for launch in early 1988. In 1990, the Canadian Radarsat should be operational, giving imagery with a resolution of between 20 and 25m. As with ERS, it will employ synthetic aperture radar, but it will be geared specifically to short-time monitoring of sea-ice conditions, especially in Arctic waters and the north-west Atlantic.

In 1986 or 1987 the Japanese propose to launch the first MOS (Marine Observation Satellite), which will deploy radiometers capable of scanning in several wavebands. It should prove a highly versatile system, combining in one payload several sensors that have been operated exclusively in earlier American and European satellites. Spatial resolutions, however, may not be more than 35-50m.

The Indian Space Research Organisation has already commissioned an imaging satellite (IRS-1) due for launch in 1986 and is planning to upgrade its programme. China is also thought to be considering an independent capability. Other nations may shortly begin to consider the advantages of space-based remote sensing systems adapted to their respective geographies. In 1984, several national and super-national agencies, including some from developing countries, joined together to form the Committee for Earth Observation Satellites (CEOS). The main objective of CEOS is to effect some degree of co-ordination and consultation for future space-based remote sensing.

The implication is that the global information 'explosion' that is already in sight will occupy a growing body of trained professional and technical personnel. The need for expansion in teaching and training facilities is urgent, otherwise the enabling technology will be vastly superior to the methodologies available to existing and potential uses of global remote sensing data.

PART 1.2: ORGANISATIONAL STRUCTURE OF REMOTE SENSING ACTIVITY IN THE UK

Remotely-sensed data relating to the British Isles has been accumulated for over 60 years. Up until the late 1960s the vast proportion of this was recorded as vertical and oblique aerial photography, mostly in black and white. Since then, there has been an expansion and diversification of types of available data as a result of the development of new optical/ mechanical techniques of image acquisition, sensor platforms and computer-based digital image processing, enhancement and analytical methods. The addition of satellite and spacecraft derived data has added new spatial and temporal dimensions, which continues to give vitality and importance to the subject. The remotely sensed data base of the country obtained by satellite surveillance has increased the value of historical sets of imagery of all types.

The organisational structure for the documentation, archiving and procurement of aerial photography, radar and infrared imagery of the land area and coastal waters of the British Isles has always been (and remains) fragmented. Although attempts at the centralised indexing of publicly available air photo cover has met with some success, it is widely acknowledged that a significant proportion of this resource remains elusive. The advent of satellite imagery has brought into being a number of archiving and documentation centres, notably the National Remote Sensing Centre. There has also been a parallel growth in other public and private organisations with various degrees of capability for receiving, processing and distributing remote sensing products. A few of these provide both aerial and space derived information, but the majority confine their attention to satellite data.

Potential users of remote sensing data are therefore confronted by a complex, decentralised and somewhat overlapping organisational structure. This situation has been clearly noted, and concisely summarised in a Report of the House of Lords Select Committee on Science and Technology, _Remote Sensing and Digital Mapping_ 1st Report, Vol 1, HMSO, 21st December 1983 (£7.00). This important assessment of the potential of remote sensing responds positively to its fundamental role as a cost- and time-effective, as well as versatile, source for the compilation and maintenance of a national geographical data base. It is recognised as a significant example of the application of information technology and a growth industry worthy of support and increased funding. The report makes a number of recommendations concerning reform of the relations between the major public bodies involved.

It is too early to detect any major response to these recommendations, but changes in the ways remotely sensed data are made accessible for public use can be anticipated.

This section will therefore attempt a brief outline of the principal public agencies involved in the provision of the raw and processed data, the development of research facilities and

LIVERPOOL INSTITUTE OF HIGHER EDUCATION THE BECK LIBRARY

other support services. The specific products produced and marketed by these bodies are listed elswhere.

A. PUBLIC AGENCIES

The National Remote Sensing Centre (NRSC)

The NRSC part of the Space Department at the Royal Aircraft Establishment (RAE), Farnborough, is the most important public agency in this field. It was established in its present form in 1980 with funding provided primarily by the Department of Trade and Industry (DTI), Overseas Development Administration, Department of the Environment, Natural Environment Research Council, Ministry of Agriculture, Fisheries and Food and the Scottish Development Department. The DTI is now (December 1985) the exclusive source of funding, and is directly responsible for the NRSC. Accommodation is provided by the Ministry of Defence, but the centre is a civilian organisation. The full address of the National Remote Sensing Centre is: The NRSC, Space Department, Royal Aircraft Establishment, Farnborough, Hants, GU14 6TD (0252 541464; Telex 858442, PE MOD 6). To give improved accessibility to the public, the NRSC moved to new premises, outside the security fence, in December 1985.

The NRSC is almost exclusively concerned with data obtained from satellites and spacecraft, although nominally it is also involved in the acquisition and dissemination of imagery from airborne sensors. The terms of reference and roles of the NRSC are:

(i) Receive, archive and supply remote sensing data and image products on a routine basis; and respond to requests for data and to develop and publicise their range and availability.

(ii) Provide facilities for the quantitative analysis of remotely-sensed data, and to support all forms of research into, and development of, processing, analytical and interpretational techniques.

(iii) Develop expertise in both the technical and applied aspects of the use of remote sensing data; engage in both 'in house' and collaborative research work and offer a comprehensive support and advisory service for existing and potential clients.

(iv) Provide or encourage the provision of educational and training facilities.

The policies of the NRSC are determined by the National Remote Sensing Programme Board, whose membership is drawn from representatives of government organisations, commercial companies, research councils and learned societies. The Programme Board is advised by the Remote Sensing Applications Committee (RSAC), chaired by the NRSC Manager. The Chairmen of the Working Groups that are supported by the NRSC are members of the RSAC. At present, there are seven Working Groups, each one of which consists of representatives of a wide spectrum of specialist interests and expertise in their respective fields. These are: Education and Training; Geological Applications; Hydrology and Water Resources; Coastal Processes; Geographic Information Systems and Information Handling Techniques; Land Applications (Land Use; Land Cover) and Oceanography/Marine Applications. The Working Groups play an important role in linking the activities of the NRSC with the broader contexts of remote sensing in the UK. They have contributed indirectly to some of the research and applications projects of the NRSC, eg the AGRISPINE project, which involved co-operation between NRSC and European agencies.

The activities and facilities of the NRSC

The NRSC accommodates the UK National Point of Contact (NPOC) with EARTHNET, one of the constituent organisation of the European Space Agency. The latter has operational responsibility for the collection, classification and general dissemination of satellite remote sensing data of all types throughout Europe. The archive of computer compatible tapes and 'browse' collections of photographic prints of satellite images are elements of this facility. In practice the 'split' between the NRSC and NPOC is of nominal significance.

The facilities for image processing and analysis provided by the NRSC range from sophisticated to 'user-friendly' systems. The latter are suitable for users with little, or no, previous experience of image processing. A full range of photographic and other facilities in support of all services is maintained.

In an attempt to make the Centre's facilities more accessible to users throughout the UK, regional sub-stations are being created, equipped with the GEMS Image Analyser and with direct access to the computer tape archive at Farnborough. The one station in operation at present is at Silsoe College near Bedford.

The publications produced by NRSC

Full details of the NRSC's many roles, facilities, products, services and recent research and application projects are presented in an attractive and informative colour brochure, and a series of Fact Sheets, both available free of charge on request. The DTI have also produced two equally attractive brochures, both with the simple title, Remote Sensing. One is devoted to an overview of the scientific and technological basis of remote sensing (together with examples of proven applications); the other is an inventory of equipment and services available from public and private organisations in the UK. Both may be obtained, free of charge, from the DTI, Space Division, 29 Bressenden Place, London SW1E 5DT (01-213 4660).

The NRSC publishes the Data User's Manual, and a quarterly Newsletter, which provides information on recent developments and forthcoming activities; the latter replaces the elaborate bi-annual RAE Remote Sensing Information Bulletin which contained a large amount of information non-specific to the

NRSC. Anyone may request to be placed on the mailing list to receive the NRSC Newsletter. The first issue was in January 1985.

B. COMMERCIAL AND QUASI-OFFICIAL ORGANISATIONS

There are a certain number of organisations that combine long experience with aerial photography and photogrammetry (principally for survey purposes) with an involvement in the use of non-photographic image products and related image processing and analysis. The major ones are all commercial companies that have developed and expanded their expertise parallel with the elaboration of remote sensing methodology over the past 20 years. By definition, their products, services and facilities are available on a commercial basis only, although they have conducted collaborative research programmes and projects with public agencies in the UK. A growing number of companies offer consultancy services exclusive to satellite and spacecraft-derived data. Collectively, these private organisations possess image archives, facilities for analysis and consultancy services of considerable significance in the UK context. Full details of these organisations are provided in later sections.

The Natural Environment Research Council (NERC)

This is an independent but government-financed public agency that makes an important contribution to the furtherance of research and applications in fields using remotely-sensed data. The NERC Scientific Services promote the application of remote sensing technology and data products in the research programmes of the organisation's various institutes (eg Institute of Terrestrial Ecology, Institute of Oceanography). Facilities and services are maintained to support this work, and also research at universities and polytechnics financed by NERC grants. These include powerful image analysis and mapping facilities, located at the Thematic Information Systems Unit (TIS), operated by the University of Reading, Department of Geography, and an aircraft dedicated to remote sensing missions of various kinds. The latter service has recently included the acquisition of high resolution aerial photographs for ecological mapping and thermal infrared scanner data. Imagery thus obtained is distributed to research groups for whom it has been obtained; the NERC does not maintain an independent image archive available for public use. (The one exception is the Meteorological Satellite Receiving Station at the University of Dundee which is NERC funded; details are given in the section devoted to meteorological satellite data.) NERC staff organise workshop meetings, usually specific to continuing research programmes, as well as training sessions for scientists and postgraduate students in receipt of research awards. A regular Remote Sensing Information Notes duplicated newsletter serves as an up-date on NERC supported activities; it also includes details of external activities that involve NERC staff. Full details of services, facilities, etc, from: NERC, Technology Information Division, Polaris House, North Star Avenue, Swindon. SN2 1EU (0793 40101). Regional requirements for image analysis facilities may be fulfilled through the BGS at Keyworth (Notts) and the ITE at Bangor (Gwynedd).

The Science and Engineering Research Council (SERC)

The potential weakness of divided responsibility was recognised by a SERC appointed committee, which reported in November 1984 in favour of exclusive responsibility falling to NERC. At the time of writing this recommendation had not been taken up, but its acceptance would at least be one positive reform of the weak coordination of remote sensing research and development in the UK.

Emerging National Co-ordination

A very encouraging note was struck by the announcement in November 1985 that the UK is to establish a British National Space Centre. Although remote sensing will be only one concern of many within this 'umbrella' organisational structure, it promises greater coordination, centralisation and improved funding. Meanwhile UK commitments to European Space Agency plans will emphasise the polar platform element of a space station programme, in co-operation with the USA.

Increasing formal liaison and co-operation between quasi-official bodies, in the interests of promoting discussion and decisions on national priorities in remote sensing is beginning to emerge. In mid 1985, the NRSC established a Land Applications Steering Group, to be chaired by the NERC, with the brief to coordinate research and applications developments, and provide advice on future programmes. This Group will probably devote much of its initial attention to microwave remote sensing.

In addition, the Royal Society has created a Remote Sensing Subcommittee of the British National Committee for Space Research (BNCSR). One of its first actions was to set up an ad hoc working group to make recommendations about optimal organisational structures needed for national/international relations in the non-government sector. The BNCSR is the UK Committee corresponding with COSPAR (Commission for Space Research), particularly the latter organisation's Interdisciplinary Scientific Committee on Space Studies of the Earth's Surface, Meteorology and Climate. COSPAR itself is ultimately regulated by ICSU (the International Council of Scientific Unions) which - at the last count - was supporting no less than 12 international working groups with terms of reference more or less directly related to remote sensing. The lack of a strong focus for remote sensing within the ICSU structure is a weakness, ultimately, for national endeavours. It is to be hoped that the Royal Society's initiative will make progress under the burden of this plethora of overlapping committees, some of whom make no mutual contact.

General Technology Systems Ltd

At present, there is coordination of interests between the NRSC Programme and DTI through the appointment of General Technology Systems Ltd to effect and promote liaison. GTS is a private, independent company, acting as impartial information and technology brokers. An important recent activity has been the completion of a survey on UK expertise oh geographical information systems. The address of the company is: Forge House, 20 Market Place, Brentford, Middlesex, TW8 8EQ (01-568 5871).

Part II

Sources of Original Imagery: Products and Services

PART 2.1: PHOTOGRAPHS AND SATELLITE DATA

A. AERIAL PHOTOGRAPHY (AND OTHER TYPES OF AERIAL IMAGERY)

Users of photographic and other image products obtained from aircraft, balloons, helicopters and other platforms operated within the atmosphere will not find a centralised archive. The numerous official, commercial and private collections of, for example, aerial photography have not been integrated and are not subject to any standardised form of documentation. All of the major repositories of airborne remote sensing data are listed in later sections. The archive maintained by the Ordnance Survey, and available for public consultation, comes closest to a national collection. The Central Register of Aerial Photography for England and Wales has recently been redistributed into constituent parts on the basis of the original ownership of prints.

A source of aerial photography that is particularly worth noting is the Aerial Photography Unit of the Agriculture Development Advisory Service (ADAS) of the Ministry of Agriculture, Fisheries and Food, which publishes an Annual Report of activities. Copies of this can be obtained from: Block C, Government Buildings, Brooklands Avenue, Cambridge, CB2 2DR (0223 358911 ext 2676).

1. Official/Public Sources

It is implicit throughout this section that collections consist of vertical aerial photos (unless otherwise stated). Until mid-1984 a Central Register of Air Photography for England was maintained by the Department of the Environment. Facilities for the indexing, inspection and provision of information on the availability of air photo cover have been transferred to the Ordnance Survey. The latter now provides an information, search and print supply service for all photography flown for the purposes of map revision since 1951. The collection is, of course, constantly updated. Full details and enquiries should be made to: Air Photo Cover Group, Ordnance Survey, Romsey Road, Maybush, Southampton, SO9 4DH.

All RAF-flown Crown Copyright air photography previously held by the Central Register has been returned to the Ministry of Defence; enquiries about coverage and services available should be made to: Ministry of Defence, F 6t 2 (Air), St George's Road, Harrogate, North Yorkshire, HG2 9DB (0423 793000).

Additional collections of wartime RAF photography of Europe (amounting to over eight million prints) is held by the University of Keele and the Pitt-Rivers Museum of the University of Oxford. The former covers mostly western and northern Europe and parts of the Mediterranean coastline, and has been catalogued. It is administered by the Department of Geography. Details of cover and access facilities are available from the Keeper of Aerial Photography, Department of Geography, University of Keele, Staffs, ST5 5BG. Requests for cover should state position to the nearest minute of longitude and latitude or a specific place name. All photos are Crown Copyright. The collection at the University of Oxford is largely confined to southern and south-eastern Europe and the margins of the Mediterranean Sea. It is less accessible for public inspection, and requests for further details should be addressed to the Museum Director, South Park, Oxford.

The former custodian of the Central Register, the Air Photographs Unit (APU) of the Department of the Environment (DoE), remains active and temporarily retains over two million prints that are to be transferred to the Royal Commission on Historical Monuments (England) where they will contribute to a new archive, the National Monuments Record. Details about this collection, as well as aerial photography specially commissioned by the DoE, are available from APU at Prince Consort House (6th Floor), Albert Embankment, London SE1 7TF (01-211 3000).

For air photo cover of Wales, enquiries should be addressed to: Central Register of Air Photographs for Wales, Welsh Office, Cathays Park, Cardiff, CF1 3NQ. For Scotland: The Air Photographs Officer, Central Register of Air Photography, Scottish Development Department, New St Andrew's House, St James's Centre, Edinburgh, EH1 3SZ (031-556 8400). An information leaflet on coverage ordering procedures and types of products is available. For Northern Ireland: Department of the Environment (NI), Ordnance Survey of Northern Ireland, 83 Ladas Drive, Belfast BT6 9FT (0232 220202).

Each of these units also have satellite image coverage of their respective areas, although the main archive is at the National Remote Sensing Centre.

Air photo collections are also commissioned and held by other national government and quasi official agencies such as the Natural Environment Research Council (which has its own aircraft, though in this case some of the imagery obtained to date has been non-photographic); Ministry of Agriculture, Fisheries and Food (Aerial Photography Unit); Forestry Commission; Soil Survey of England and Wales; Transport and Research Laboratory; Nature Conservancy Council (Remote Sensing Unit, Edinburgh), British Geological Survey and many others. It is often difficult to identify the location and dates of coverage, as indexing and retrieval varies in efficiency. Most County and Metropolitan District Planning; Technical Services; Engineering and related departments; National Park and Water Authorities hold air photo collections, some of which have been copied from Ordnance Survey originals. However, significant collections of often high quality air photos have been commissioned by many local authorities for specific planning projects or for routine monitoring (e.g. of coastal changes). There is no centralised documentation of this potentially valuable cover and enquiries need be addressed directly to the authority offices in question. From time to time, these collections are rationalised and unwanted material is often sold off. Teachers looking for local or regional cover could find this to be a possible source of air photos, which are normally too expensive for limited school budgets. The addresses of all organisations are to be found in Environmental

Education: Sources of Information 1981, Department of Education and Science, HMSO (new edition in preparation).

2. Commercial and Institutional Sources

An outstanding collection of many hundreds of thousands of both vertical and oblique air photos of the British Isles is maintained by the Committee for Aerial Photography, University of Cambridge, Mono Building, Free School Lane, Cambridge CB2 3RF. This is an expertly indexed archive of special interest to historians, historical geographers and archaeologists.

Most university and polytechnic departments (especially geography, geology, land survey and civil engineering) hold relatively small collections of aerial imagery, which are usually biased towards local area cover. In most cases, imagery is derived from the primary sources already noted, but there may be useful collections of cover flown for specific research projects.

Commercial companies specialising in aerial surveys and in remote sensing consultancy work have acquired large collections of original air photographs and other types of imagery obtained using airborne sensors (e.g. thermal linescan; side-looking airborne radar). It is usually subject to the copyright of clients for whom the cover was originally obtained. In most cases, coverage of overseas areas is at least as strongly represented as that of the UK. Enquiries specifying area(s) of interest, scales, date(s), etc, should be addressed in the first instance to the Photo Library. Potential users will usually need to visit company offices to make direct inspection of imagery. The major companies are:

(i) Clyde Surveys Ltd, Reform Road, Maidenhead, Berks, SL6 8BU (0628 21371).

(ii) Nigel Press Associates Ltd, Old Station Yard, Marlpit Hill, Edenbridge, Kent, TN8 5AV (0732 865023).

(iii) Hunting Surveys Ltd, Elstree Way, Borehamwood, Herts, WD6 1SB (01-953 6161).

(iv) BKS Surveys Ltd, Ballycairn Road, Coleraine, County Londonderry, Northern Ireland, BT51 3HZ (0265 52311) (with subsidiary branches in over 15 overseas centres).

(v) Meridian Air Maps, Marlborough Road, Lancing, Sussex, BN15 8TT (0903 752992).

(vi) Geosurvey International Ltd, Geosurvey House, Orchard Lane, East Molesey, Surrey, KT8 0BY (01-398 8371/2).

There are in addition a number of smaller companies with essentially local or regional spheres of operation. Many specialise in oblique rather than vertical air photos and service the requirements of publishers. Their names are most conveniently located under 'Aerial Photography' entries in local telephone directories and Yellow Pages.

Both national and provincial newspapers maintain files of mostly oblique air photos, some of them dating back several decades, which can be copied.

The outstanding collection of high quality oblique aerial photography of all parts of the British Isles is maintained (and frequently updated) by Aerofilms Ltd. It is summarised in The Aerofilms Book Of Aerial Photographs (1972), containing over 500 black and white photographs; despite its age, this remains a structured and essential guide to this extensive library. Block vertical aerial photography of several parts of England and Wales is also held, together with a selection of photo mosaics (e.g. of Central and Greater London at 1:10560). Many of the outstanding oblique views (and some vertical stereo pairs) have been combined into a wide range of 'Educational Sets' of prints and slides. (These are discussed in detail in the Resources for Teachers section.) The emphasis of Aerofilms' approach is to satisfy individual requirements and interests. Paper copies of prints will be sent to enquirers before commitment to purchase. For many areas, the photographic record extends back 40-60 years. Organisations and individuals may also commission aerial photographic cover of specified areas. Full details of products, services and costs on request. Aerofilms Limited, Gate Studios, Station Road, Borehamwood, Herts, WD6 1BJ.

3. Sources of Overseas Coverage

There is insufficient space in this publication to attempt a summary of the availability of aerial photography in countries outside the UK. There are well-documented central registers or libraries in certain countries such as USA, Canada and France, and scattered collections under various restrictions of copyright or security in many others. Anyone wishing to acquire air photo cover of developing countries, without prior knowledge of their provenance, should contact the Remote Sensing Unit of the Food and Agriculture Organization (FAO) in Rome, Italy. The FAO has taken on the formidable task of compiling a world index of remote sensing imagery and will at least be able to advise on potentially helpful national and regional agencies. The library of the Royal Geographical Survey (1, Kensington Gore, London SW1) may also be able to offer advice, based on a survey conducted in 1972.

Considerable collections of aerial photography and other types of aerial imagery of various overseas countries are held by British remote sensing and air survey consultancy companies such as Huntings, Clyde Surveys, BKS Surveys, Hunting Surveys, Meridian Air Maps and by government organisations such as the Directorate of Overseas Surveys, a constituent part of the Ordnance Survey (Technical Support Services Branch, Directorate of Overseas Surveys, Ordnance Survey, Romsey Road, Maybush, Southampton, SO9 4DH). Initial enquiries about availability should specify geographical area of interest and sortie date(s); inspection of positive prints in the libraries of these organisations is the next step. Large sections of all of these collections are the copyright of originating clients, and clearance of any restrictions thus imposed is a prerequisite to any order for copies. The Royal Signals and Radar

Establishment, Malvern, Worcestershire has acquired coverage of different types of radar imagery, but this is not a publicly accessible collection or archive.

In the United States, the EROS Data Center maintains an archive of US Geological Survey and NASA sponsored aerial photography, which is publicly available. An invaluable reference source is Sources of Information and Materials: Maps and Aerial Photographs, published by the Association of American Geographers, Washington, DC, USA. Further sources are the US Department of Agriculture, Aerial Photography Field Office, PO Box 30010, Salt Lake City, Utah, 84125 USA and the National Cartographic Information Center, 507 National Center, Reston, Virginia 22092, USA. Further details on US air photo collections and archives are given in Everyone's Space Handbook, D. Kroeck (1976), Pilot Rock Inc, USA.

Canadian air photo coverage is efficiently accessed via the National Air Photo Library (Department of Energy, Mines and Resources), 615, Booth Street, Ottawa, K1A OE9, Canada which also provides contact with provincial collections. The NAPL provides a concise but very helpful booklet How to Order Aerial Photographs. It is a model of its kind.

There are large, well documented collections of aerial imagery of France and both past and present overseas departments and territories held by the Institute Geographique National, Centre de Documentation de Photographies Aeriennes, 2 Avenue Pasteur, 94160 St Mandé, France.

B. NON-METEOROLOGICAL SATELLITE AND SPACECRAFT IMAGERY

The National Remote Sensing Centre

User services at the National Remote Sensing Centre (NRSC), Space Department, Royal Aircraft Establishment, Farnborough, GU14 6TD (0252 541464), include film writing and digitising data received via Earthnet and from other data centres; interactive digital image processing and digital mapping; internal and co-operative research into the interpretation and application of all forms of satellite data and the maintenance of an archive and 'browse facility' of hard copy black and white photographic prints of Landsat and other satellite imagery of the British Isles (see Addenda). A 'quick look' automatic facsimile image transmission link with Earthnet now extends this facility to coverage of the entire European area. The browse facility is located in a library equipped with other reference material such as Ordnance Survey maps, atlases, microfiche catalogues of worldwide Landsat, Seasat, MOMS and MC data, as well as a selection of books, monographs and reports on remote sensing. The NRSC has produced a set of 24 Fact Sheets describing its products, services and facilities, as well as general remote sensing background details. These are available on request, and provide an attractive summary of the functions and activities of the NRSC. All types of potential users of remote sensing data are encouraged to contact the centre.

The staff of NRSC include specialists in data processing and a team of application scientists. One role of the latter is to advise and assist customers in appropriate methods and approaches relevant to their field(s) of interest. A mobile exhibition facility equipped to demonstrate and display characteristic products and interactive image analysis (and generally indicate the potential of remote sensing) is available for visits to conferences, seminars and other meetings where condsideration of remote sensing is directly or indirectly relevant. It may be booked for this purpose by contacting the Publicity and Promotions Section, Department of Trade and Industry, Millbank Tower, London, SWIP 4QU.

Although the focus of interest is on satellite data, digitising of aerial photography can be undertaken, especially relevant to studies of environmental change over time-scales extending back to before the advent of space-based remote sensing.

The NRSC has produced a number of general interest and educational products, which are listed separately in the relevant section of this guide. Its 'menu' of photographic and digital products is given below, with respect to the following satellite imagery sources: Landsat MSS, Landsat TM, Coastal Zone Colour Scanner (CZCS) (1978-1984); HCMM (1978-1980); Seasat (1978); Meteosat (selective archive) and NOAA Polar Orbiting Satellites (also selective). (Full details are given in the Data User's Manual)

1. Photo Products

Black and white

Code	Size	Type
PP240/1	240 mm	Print
PP240/E	240 mm	Print
PP480/2	480 mm	Print
PP480/E	480 mm	Print
PP960/4	960 mm	Print
PP960/E	960 mm	Print
FN240/M	240 mm	Film negative
FP240/M	240 mm	Film negative

240 mm = scale of 1:1,000,000
480 mm = scale of 1:500,000
960 mm = scale of 1:250,000

Colour (Standard Colour Composites)

Code	Size	Type
CPP240/1	240 mm	Print
CPP240/E	240 mm	Print
CPP480/2	480 mm	Print
CPP480/E	480 mm	Print
CPP960/4	960 mm	Print
CPP960/E	960 mm	Print
CFP240	240 mm	Positive
CFN240	240 mm	Negative

Non-standard enlargements or extracts of standard images, and transparencies of photographic prints can also be supplied.

2. Digital Products

Code	Description
CCT01	Copy of archived tape (in received format)
CCT02	Copy of archived tape
CCT03	Copy of archived tape
CCT05	UK 100 metre geo transformed (from CCT01)
CCT06	UK 50 metre geo transformed (from CCT01)

(CCT = computer compatible tape)

There are advanced plans to acquire and process SIR-B and SPOT data.

3. Services

(i) Geometric correction of Landsat images to specified pixel sizes and map projections; digital mosaics.

(ii) IDP 3000 interactive image processing system.

(iii) GEMS interactive image processing system and GEMSTONE algorithms for image analysis. Both of the above can be used to achieve multitemporal analysis of sequences of geometrically transformed scenes of given locations.

(iv) LS-10 interactive image processing system (microcomputer).

(v) 35-mm colour slides from IDP/GEMS (eg for teaching or publication).

(vi) 5" x 4" colour negatives from IDP/GEMS (photo products of enhanced and classified images).

Professional help and guidance in the use of analytical equipment is provided, when required. The GEMS facility requires only basic tuition and no special knowledge of computer programming. The LS-10 is ideal for student demonstration and use.

A regional centre of the NRSC has recently been established at Silsoe College (Bedford, MK45 4DT, 0525 60428), where the Gems image analysis system is available for the interpretation of satellite imagery by research organisations, commercial companies, etc, using a suite of analytical programmes. Several CCTs are held as a 'core' library, and others that are not available can be obtained via the NRSC. A further, more limited, outstation has been set up at the Remote Sensing Unit, Macaulay Institute for Soil Research, Craigiebuckler, Aberdeen, AB9 2QJ (0224 38611, ext 243).

Although the emphasis of the NRSC is on coverage of the British Isles, the Centre's involvement in project and consultancy work in other parts of the world has resulted in a number of CCT's and photographic products for locations 'out of area'. This component of the archive is being continuously added to through requests from users and by tape exchanges. The Centre is able to obtain image data for any location in the world, given precise specifications. It may, however, be advantageous for potential users to apply to other databases directly. Some are listed below, with advice on how to make contact and find out details of the products and services offered.

European Agencies

The National Remote Sensing Centre is only one component of the Space Department of RAE, Farnborough. Another is the National Point of Contact (NPOC) with EARTHNET, an organisation within the European Space Agency. EARTHNET has responsibility for collection and dissemination of a wide variety of different types of satellite remote sensing data throughout Western Europe. It operates its own, and uses other existing national, facilities for the reception and transmission of data for NPOC archives and eventual users. Thus, Landsat data is received directly at Maspalomas (Canary Islands), Spain; Fucino, Italy; and Kiruna, Sweden; Seasat data was received at RAE Oakhanger; HCMM (Heat Capacity Mapping Mission) data at Lannion in France; and so on. The further example of Meteosat data is given later. The UK NPOC thus has direct access to an extensive satellite remote sensing data base, which may be activated by any user working in liaison with the NPOC or NRSC. The most recent acquisition is 'metric camera' (MC) black and white and colour infrared photography at a scale of 1:50,000, which was operated by the ESA Spacelab on board the Space Shuttle mission in late 1983. A catalogue of image frames and scene microfiche are available at the NRSC; prints can be obtained via the NPOC through EARTHNET. EARTHNET in turn has access to other international data bases. Imagery obtained on experimental and evaluation flights, e.g. SAR 580 aircraft campaign (1981) is also available through EARTHNET. Standard EARTHNET products and their prices are available on request via ESRIN (address overleaf). All requests should be addressed to 'EARTHNET User Services'.

Requests for data should be accompanied by sufficiently detailed information to accurately identify area(s) of interest, particularly latitude and longitude co-ordinates. A map, with area or location clearly marked, is a sufficient alternative. Indexes of European cover are published by EARTHNET and are available via the NPOC. The NRSC is at present compiling world maps of its Landsat holding to facilitate reference, consisting of three digit path and row numbers. Using the reference system, users can request a listing of all available imagery, which includes an evaluation of quality. A specific choice can then be made without the need to consult directly the browse file or microfiche copies of photographic prints. A similar service is offered for worldwide Landsat imagery by the EROS Data Center. SPOT-Image will offer an indexing and retrieval service once SPOT-1 has been launched in late 1985 (see p. 56).

The European Space Agency (ESA) operates an Information Retrieval Service (ESA-IRS) as part of the EARTHNET Management Centre at Frascati, Italy (ESRIN, via galileo galilei, CP 64 00044, Frascati, Italy, (06) 94011). This facility is the interface between the reception, pre-processing and archiving of data on the one hand and the interests of all data users on the other. The status of the EARTHNET archive for any part of Europe and North Africa can be ascertained via IRS, using the LEDA-2 (On-Line Earthnet Data Availability) which is a direct on-line interrogation and retrieval link, and covers a variety of satellite imagery types. Users can search these files and place orders for products (tapes or pictures) either directly or through national NPOC's RETROSPECT and LEDA CURRENT files, according to requirements. For data of specific areas it is often advantageous to contact the relevant national NPOC. This is because the latter may have generated products for previous users that are not recorded on LEDA.

ESA also publishes Earthnet Review, a magazine published twice a year giving information on recent and future developments, and catalogues of all image acquisitions, which are updated by frequent supplements. It should be noted that some of these products are for remote sensing missions that are no longer operative such as Seasat and Nimbus-7 CZCS (Coastal Zone Colour Scanner). Users with a particular interest in Seasat would be advised to contact the UK NPOC (Farnborough) directly. General enquiries for data distribution, products, etc should be addressed to ESA, 8-10 Rue Mario Nikis, 75738, Paris 15, France, or EARTHNET (address above).

LIST OF EUROPEAN NATIONAL POINTS OF CONTACT (NPOC)

Belgium

Services du Premier Ministre
8 rue de la science
1040 Bruxelles
02-511 59 85

West Germany

DFVLR Hauptabteillung
Raumflugtrieb
8031 Oberpfaffenhofen
Post Wessling
08-153 281

Denmark

National Technological Library
Centre for Documentation
Anker Engelundsvej 1
2800 Lyngby
02-88 30 88

Ireland

National Board for Science and Technology
Shelbourne House
Shelbourne Road
Dublin 4
01-601 797

Netherlands

National Aerospace Laboratory
Anthony Fokkerweg 2
Amsterdam 1071 NL
020-511 31 13 ext 291

Italy

Telespazio
Corso d'Italia 43
00198 Rome
06 8497

France

GDTA
Centre d'Etudes Spatiales de Toulouse
18 Ave Edouard Belin
31055 Toulouse
61-53 11 12

Spain

CONIE, Pintor Rosales 34
Madrid 8
01-247 98 00

Sweden

Satimage
PO Box 816
28 Kiruna
Kir. S-981

ESA-IRS offers a 'Basic Search Library' (BASLIB) service that identifies bibliographic citations on all types of publications in specified fields. One of these is Earth Observation, divided into: infrared imaging; pattern recognition; photo and infrared detectors and signal detection and processing. For a subscription, BASLIB will provide a print-out of all citations, at a monthly frequency. Potential users in the UK should contact the IRS national centre: IRS-DIATECH, Room 204, Ebury Bridge House, 2-18 Ebury Bridge Road, London, SW1W 8QD.

ESA also publishes Earth Observation Quarterly, which provides up to date information and special features on current developments concerning remote sensing and space science in Europe. It is written in a direct style and is available free of charge. Requests to be placed on the distribution list should be addressed to ESA Scientific and Technical Publications Branch, ESTEC, Postbus 299, 2200 AG Noordwijk, Netherlands. The ESA Journal (first published in 1976) is also published quarterly and contains substantial articles recording the results of research conducted within ESA or within European agencies and institutions. It is free of charge to professionally-qualified applicants. Numerous special publications, technical reports, notes and memoranda, and brochures have been published, including a specialised series of Procedures, Standards and Specifications. A catalogue of all titles is available on request from the above address; microfiche copies of out of print titles may be ordered via the

French office of ESA-IRS. Only a minority of these various publications are specific to remote sensing.

Academic, Institutional and Commercial Organisations

Several academic and commercial organisations also house collections of satellite imagery, in both photographic print and digital formats. Most are equipped for data analysis at various levels of complexity. Most universities and polytechnics with an involvement in remote sensing hold collections that service teaching requirements and reflect past and continuing research projects. In the majority of cases, satellite imagery is catalogued with centralised or departmental collections of maps and air photos. The institutions with the largest collections are those listed in Part 3.1. A detailed inventory of both imagery holdings and equipment was given in the Department of Trade and Industry's Remote Sensing of Earth Resources (1981, 4th Edition). This is obviously now in need of up-dating, especially with respect to image processing facilities.

A number of universities have established opportunities for consultancy work with outside clients (see later section). There are also a number of commercial companies that specialise in the archiving and processing of satellite data on behalf of clients for whom they act as expert consultants. Their holdings reflect involvement in project investigations and surveys in a variety of locations worldwide. These collections are accessible to individual users. The principal sources are given below.

1. Clyde Surveys Ltd (Environment and Resources Consultancy)

Their Landsat Products Service includes all aspects of acquisition, processing and interpretation of different Landsat image formats. Final products are provided in the form of prints, slides and CCT's. 'In house' training on most aspects of image processing and interpretation is available to employees of client organisations (at the latter's premises, if necessary). The library of Landsat images is particularly strong on most parts of Africa, the Middle East and South-East Asia and is subject to continuous up-dating and addition. As previously noted, Clyde Surveys also hold a large collection of aerial photography of both the UK and overseas which can be interfaced with satellite-derived imagery.

Requests for further details, and specific coverage to: Clyde Surveys Ltd, Landsat Products Service, Reform Road, Maidenhead, Berks, SL6 8BU (0628 21371).

2. Geosurvey International Ltd

This company has developed its own 'in house' digital image processing system to handle both satellites and geophysical data, as well as a specific technique to provide geometrically rectified image maps for resource and other surveys at regional scales of investigation. These facilities are largely dedicated to product development for clients. Geosurvey also has extensive experience in the production of maps from aerial photography. The company is unique in the UK in having high altitude jet aircraft in its survey fleet. (Geosurvey Ltd, Orchard Lane, East Molesey, Surrey, KT8 OBY; 01-398 8371/2).

3. Hunting Surveys Ltd

Although primarily a company offering a full service for resource inventory and appraisal, project implementation and monitoring, and survey planning and analysis, their employment of all types of remote sensing data and analytical techniques has generated a large collection of imagery which is available for public inspection via the company's library. Coverage includes a wide variety of world locations, including the UK, Ireland, north-west Europe, India and Pakistan, United Arab Emirates, south-eastern Africa, Madagascar, Bolivia, Panama and other areas of Latin America, Algeria and Sudan. The majority holding is that of Landsat imagery, but Seasat SAR, SIR (Shuttle Imaging Radar) scenes and mosaics are also of considerable interest. The latter represents one of the most accessible large collections of radar imagery from space in the UK, outside of the holdings of the NRSC.

Further details are available from the company at Elstree Way, Borehamwood, Herts, WD6 1SB (01-953 6161). The three constituent parts are: Hunting Geology and Geophysics, Hunting Surveys and Hunting Technical Services Ltd.

4. The British Library: Map Library

The Map Library of the British Library accommodates copies of all major texts, monographs and journals in remote sensing (as well as texts in the earth, environmental and biological sciences). On-line searches of computer data bases (e.g. Geo Archive) for bibliographical citations are undertaken by arrangement or appointment, and an immediate or postal photocopying service is also operated. The Map Library division is specifically responsible for maintaining an archive collection of various types of satellite imagery of the British Isles, and catalogue of NRSC image archives, e.g. Seasat coverage of the Earthnet area. In addition, the Landsat Microcatalog, published by the EROS Data Center, is available for consultation, thus providing worldwide coverage of imagery dating back to 1972; Landsat, and Skylab, imagery may also be viewed as 16-mm browse film. Work on cataloguing other remote sensing data bases of the UK is in active preparation. This reference facility is now of major importance to all potential users in the UK especially in view of its public accessibility. Further details from: The British Map Library, Great Russell Street, London WC1B 3DG (01-636 1544 ext 265). It is open to visitors, Monday to Saturday, from 10 am to 4.30 pm, except for occasional public holidays.

5. Nigel Press Associates Ltd (NPA)

NPA has a similar scope as Clyde Surveys, though has the longest experience of archiving and processing satellite data of any British company. The number and variety of projects and surveys carried out since 1972, some of them on continent-wide scales, has contributed one of the largest independent collections of satellite imagery anywhere in the world. Landsat data dominates, but there are also holdings of earlier Skylab, Apollo and Gemini photographs and imagery from other satellite systems on a selective basis. SPOT imagery will be added, once it is available. The majority of the company's consultancy projects have been in the geological field, although there has been more recent diversification into application areas such as forestry, water resources and hydrology, agriculture and environmental pollution. NPA maintains an Earthsat Data Centre and can provide index diagrams of their Landsat holdings for any part of the world. This includes mosaics of extensive parts of Arabia and Africa. Search and acquisition of imagery from data bases such as the EROS Data Center and individual Landsat receiving stations is undertaken on behalf of clients. Various routines for enlargement, enhancement and analysis of image holdings are available. The EDC coverage is complete for Europe, Africa and Arabia, and extends to Latin America and southern and south-east Asia.

Teachers will be attracted by the provision, free of charge, of paper copies of image prints for preview. NPA also offers collections of older stocks of photographic prints at very low prices on an occasional basis. This is a very cost-effective way of acquiring teaching collections or building up a basic collection of satellite imagery for demonstration, discussion and display. Slides of single images can also be provided. NPA has shown a particular concern for the low budget constraints of schools, colleges and higher educational establishments, and their prices are as low as they can be for educational institutions. Special teaching collections (to individual specifications) of ten Landsat prints of zonal land forms and ecosystems are available at low prices; a further teaching set is available to show Landsat imaging wavelengths and image scales.

NPA (under the name of CARTOSAT) has specialised in the production of image mosaics, and 'image-maps' which add place-name annotation, a geometric co-ordinate reference base, contours and other data to original imagery. These have been produced as 'one-off' compilations for clients, but are subject to copyright restriction and copies are available for general purchase.

Further details and fully-illustrated publicity brochures on all aspects of the activities of this company from: Nigel Press Associates Ltd, Lansing Building, Old Station Yard, Marlpit Hill, Edenbridge, Kent, TN6 5AU (0732 8865023).

6. Environmental Remote Sensing Applications Consultants Ltd (ERSAC)

Established in 1983 as an applications-oriented remote sensing organisation, with international as well as specifically Scottish interests, ERSAC offers a fully-equipped consultancy service particularly in the field of the earth and environmental sciences. This has involved the development of a data exchange facility in the effort to create an international satellite computer tape archive. ERSAC is therefore an important centre for the existing and potential user community in Scotland and the north of England, as well as in the UK as a whole.

A comprehensive descriptive brochure of the company's services, facilities and products, including an account of recent projects is available from: ERSAC Ltd, Peel House, Livingston, Scotland, EH54 6AG (0506 412000).

7. Other UK Sources

The British Library (Map Library) has recently commissioned the compilation of a descriptive catalogue of existing remote sensing and cartographic data bases and data sets in the UK. This will be published and used as a basis for computer searches for information. Data bases for overseas areas created in the UK will also be included. The work is being carried out by the Department of Geography, Birkbeck College, University of London. It should be operational by late 1985.

The Scottish Development Department (Air Photographs Unit, New St Andrews House, St James Centre, Edinburgh EH1 3S2) also holds catalogues and microfilm copies of Landsat imagery of Europe and certain other areas, which may be viewed by appointment.

If satellite imagery for areas outside of Europe is required, and cannot be provided by any of the UK sources listed above, the following archives can be consulted. (The main sources of meteorological data are given in the next section.)

8. EROS Data Center, USA

The EROS Data Center is a part of the Earth Resources Observation System (EROS) Program of the Department of the Interior, managed by the US Geological Survey and Department of Commerce. It is the national centre for the processing and dissemination of spacecraft and aircraft-acquired photographic imagery of the Earth's resources (see Addenda).

The EROS Data Center provides access to Landsat data, aerial photography acquired by the US Department of the Interior, and photography and other remotely sensed data acquired by the National Aeronautics and Space Administration (NASA) from research aircraft and from Skylab, Apollo, Gemini and Shuttle spacecraft (large format camera imagery, 1984, in the latter case).

Orders for photos and images, enquiries on the availability of coverage over specific areas, and requests for price information should be directed to: User Services, EROS Data Center, Sioux Falls, South Dakota, 57198, USA (605-594 6511).

At the heart of the Data Center is a central computer complex that controls a data base of more than ten million images and photographs of the Earth's surface, performs searches of specific geographic areas of interest, and serves as a management tool for the entire data reproduction process. The computerised data and retrieval system is based on a geographic system of latitude and longitude, supplemented by information about image quality, cloud cover, and type of data. Guided by customer requirements, a computer geographic search will print out a listing of available imagery and photography from which the requester can make a final selection.

EROS is also able to supply NASA-prepared US Standard and Non-US Standard Catalogs (which are cumulative) of Landsat imagery. These are directly available in the UK from NTIS (UK Service Centre), PO Box 3, Alton, Hants, GU34 2PG. Since 1983, a Landsat MICROCATALOG, offering a comprehensive microfiche reference system for all Landsat data, has been developed for users without ready access to EROS data bases.* Both are listings of image details. Indexes of data from different satellites, produced by the National Mapping Division of the USGS are also supplied.

Potential users should request a copy of the detailed explanatory booklet which gives specific details of the archive format and search, procurement and ordering information. Applications for geographical search and listing of image holdings need to be entered on official forms. Once received, a computer print-out of the EROS archive relevant to specified location(s) will be provided, free of charge. A catalogue of all products and services, together with costs will be included automatically with every request for information or assistance. (Customers are required to set up a payment account or send a prepayment with orders.)

All general enquiries about Landsat-specific products should be addressed to NOAA Landsat Customer Services at the EROS Data Center. NOAA (National Oceanic and Atmospheric Administration) has responsibility for the commercial functions of EROS and all other US public agencies with involvement in data reception, archiving, processing, research and development and other technical roles in civilian remote sensing.

The EROS Data Center and NASA have published a substantial manual on all aspects of the Landsat programme and its characteristic products titled Landsat Data Users Handbook. The

---

*Details of this system should be requested from NOAA/NESDIS Landsat Customer Services, Mundt Federal Building, Sioux Falls, South Dakota 57198, USA, which provides an explanatory booklet on the format and ordering arrangements. The same office also publishes the Landsat Worldwide Reference System Index (WRSI) as a series of large, folding maps. These are currently available free of charge.

fourth edition (1984, $16.00) is available from Distribution Branch, Text Products Section, USGS, 604 South Pickett Street, Alexandria, Virginia 22304, USA. Updating on the Landsat programme is provided by Landsat Data Users Notes, a magazine-style quarterly publication circulated free of charge by NOAA. Requests to be placed on the mailing list should be addressed to NOAA Landsat Customer Services. Information of a more general nature (e.g. forthcoming conferences, new publications, other satellite programmes and projects) is also included, but the importance of this valuable publication is timely and detailed information on changes in the status of Landsat.

Not all Landsat imagery is recorded on file at the EROS Data Center. There are several Landsat data reception and processing stations operated by government agencies and other organisations worldwide. These stations may have more extensive and/or more recent coverage of areas within their reception areas (see Fig 5, Chapter 1), and should be contacted by users with relevant research or teaching interests. Included are the following (EARTHNET stations are noted on p. 47-48).

Country	Enquiries to:
Argentina	Comision Nacional de Investigaciones Espaciales Centro de Prosesamiento Dorreo 4010 1425 Buenos Aires
Australia	Australian Landsat Station Data Processing Facility 14-16 Oakley Court PO Box 28 Belconnen ACT 2616
Brazil	Instituto De Pesquisas Espaciais (INPE) Departmento de Producas de Imagens ATUS - Banco de Imagens Terresties Rodoua Presidente Dutra KM210 Cachoeira Paulista CEP-12-630
Canada	Integrated Satellite Information PO Box 1150 Prince Albert Saskatchewan Canada S6V 5S7

or

Canada Centre for Remote Sensing
717 Belfast Road
Ottawa
Ontario
K1A 0Y7

<u>China</u> — Academia Sinica
Landsat Station
Beijing

<u>India</u> — National Remote Sensing Agency
4 Sardar Patel Road
Hyderabad 500-003
Andhra Pradesh

<u>Indonesia</u> — Bakosurtanae
Jln Raya
Jakarta Bogor 46

<u>Japan</u> — Remote Sensing Technology Center of Japan
7-15-7 Uni-Roppongi Building
Miriatoku
Tokyo 106

<u>South Africa</u> — National Institute for Telecommunications Research
Satellite Remote Sensing Centre
PO Box 3718
Johannesburg 2000

<u>Thailand</u> — Remote Sensing Division
National Research Council
Bangkok 9

Each centre produces its own catalogues, in various formats, and is under contract with NASA (the supervising agency) to respond to user requests from whatever source. Other regional centres have been created, and more may develop in the future.

Skylab imagery (1973-74) is listed comprehensively in <u>Skylab Earth Resources Data Catalog</u>, published by the US Government Printing Office, Washington DC, 20402 at $12.50 (185pp, including many photographs, Publication No. 3300-00586, 1974). It provides an index to 35,000 scenes taken by the Skylab Earth Resources Experiment Package and gives a description of the remote sensing programme undertaken by Skylab crew missions. A microfilm format of Skylab imagery can be consulted at the NRSC and the British Map Library.

Skylab photographs are held by Nigel Press Associates. A complete archive, which also includes earlier Gemini and Apollo mission images, is maintained by Technology Applications Center (TAC), University of New Mexico, Albuquerque, New Mexico, 87131, USA. TAC also produces other publications (some of them listed elsewhere) and maintains an archive of US Landsat scenes.

9. Other Data Bases

The World Bank published a 17 page Landsat Index Atlas in 1975 which displays satellite image coverage for each continental land mass on 14 maps. It is a pity that this very useful publication has not been revised. Available from: Johns Hopkins University Press, Baltimore, Maryland, 21218, USA. The World Bank also makes Landsat images and derived thematic maps available through its US map distribution agent International Mapping Unlimited (IMU), 4343 39th Street, NW, Washington DC, 20016. Mosaics and country Landsat maps at scales of between 1:1 million and 1:100,000 are available for Nepal, Bhutan, Bangladesh and parts of India, Burma, Pakistan and other Asian states. (An up to date listing of all available mosaics and sheets is given in Part 2.4).

The task of producing a comprehensive world index of satellite imagery, aerial photography and other aerial imagery (e.g. radar) has been addressed by the Remote Sensing Centre of the Food and Agriculture Organisation (FAO), Rome. A group of advisory experts commenced work in 1979, and decided to give priority to developing countries, especially in Africa. The problems involved, and progress recorded, are summarised by C.I.M. O'Brien, in the Proceedings of the 10th Conference of the Remote Sensing Society (Reading, UK, 1984, pp. 313-319).

Imagery obtained from Space Shuttle Imaging Radar (SIR-A)(SIR-B) remote sensing missions is catalogued and distributed by the National Space Science Data Center (OSTA Data), NASA, Goddard Space Flight Center, Greenbelt, Maryland, 20071, USA and (more selectively) by the European Space Agency, 8-10, Rue Mario Nikis, 75738, Paris 15, France. All of the original film exposed by the large format camera (LFC) operated from the Space Shuttle mission 41G in October 1984 is held by the EROS Data Center. Accession aids such as microfilm of LFC images, microfiche catalogues and world coverage maps are available to assist users. The purpose of the LFC experiment was to obtain high spatial resolution, stereo imagery suitable for precision cartography from nominal altitudes of between 240 and 370km. Although coverage achieved was less than anticipated, the quality of the imagery obtained has been described by researchers in superlative terms. The principal distribution centre for HCMM data is World Data Center (A) for Rockets and Satellites, Code 601.4, Goddard Space Flight Center (address as above).

At the time of writing, the remote sensing community is keenly awaiting the launch of the French SPOT satellite, which will provide data with a spatial resolution of between 10 and 20m. There will also be a number of other innovative features compared to Landsat, such as off-nadir viewing. The organisational structure for the reception, archiving and distribution of SPOT data has already been established and is described in a fully illustrated brochure produced by Centre National d'Etudes Spatiales. This may be obtained from the company responsible for the programme, SPOT-IMAGE, 16 bis, avenue Edouard Belin, F31055 Toulouse, Cedex, France. A further booklet on the technical components of the satellite and its orbital configuration is also published free of charge by SPOT-IMAGE. A twice-yearly bilingual SPOT Newsletter is distributed without charge to bona fide requesters, and gives updating on all recent and current developments. It should be made clear that the SPOT project is outside of the aegis of the European Space Agency, although it has already been agreed that the NRSC and NPA will serve as accredited agencies for SPOT image product distribution in the UK, albeit on the basis of selective acquisition. It represents the first opportunity to use a system designed, constructed, launched and operated independent of US facilities and services. (The first ESA satellite, due for launch in 1988, will reinforce this operational shift.) The NRSC, in co-operation with Hunting Surveys and the NERC, commissioned aerial simulated SPOT imagery over test sites in the UK during 1982-83 (see Addenda).

Another independent European intiative is the MOMS system (Modular Optoelectronic Multispectral Scanner) developed under contract by the West German Aerospace Research Establishment (DFVLR) and tested on NASA Space Shuttle STS-7 and STS-11 flights in 1983 and 1984. It is designed to acquire high resolution data for specific purposes on relatively short missions. It can therefore be deployed for a few days on a Shuttle platform or operated independently in orbit for perhaps several months between launch and retrieval by separate Space Shuttle missions. The technical specifications of this system, including the data acquisition and information network concepts that will be applied, are given in a boldly-presented, informative booklet published by DFVLR. This gives full colour illustrations of imagery obtained from pre-flight simulations and during the test missions. A catalogue of data products is available as a separate publication from: DFVLR, D-8031 Oberpfaffenhofen, Federal Republic of Germany.

C. METEOROLOGICAL SATELLITE DATA AND RELATED PRODUCTS

Introduction

Since the pioneering launch and operation of TIROS-1 in 1960 by NASA, over 30 satellites dedicated to the acquisition of meteorological data have been placed in orbit. They have performed useful lives of an average of five years and collectively they present the most obvious, immediate benefit of operational remote sensing. The main families or programmes of meteorological satellites are noted in the introductory chapter. Together, they collect a wide range of environmental data, such as surface temperatures, atmospheric humidity, water-ice boundary definition, heat budgets, ozone distribution as well as patterns of cloud cover. Very little of this type of data is used in routine weather forecasts presented to the public, but it is selectively archived as a repository of basic data that has already made a fundamental contribution to improved descriptive and explanatory models of the Earth's atmospheric system. The short-lived usefulness of meteorological data thus becomes beneficial over a much longer period to the science of climatology. Satellites are not independent of other types of remote sensing of the atmosphere (e.g. by radio-sonde balloons) or of ground-based techniques, such as the recently implemented system of radar tracking of individual rain storms. However, satellites can also 'interrogate' automatic data collection platforms, such as buoys and ships. Together they have contributed significantly to timely responses to rapidly developing, large magnitude weather events (from hurricanes to short intensive storms), thus aiding public protection and reducing marginal losses. Information on current weather is not only transmitted to national forecasting services, but to civilian and military personnel for whom it may be quantitatively and qualitatively vital to decision-making.

Apart from their nominal value to forecasting and research in the atmospheric sciences, data from several meteorological satellites have found useful applications in other tasks of regional and global environmental monitoring. Some of these have made a direct impact on the management of critical situations, e.g. oil pollution spills in coastal waters and shallow seas, forest fires and stubble burning. Because of the coarse spatial resolution of data from these systems, only events or changes covering large areas can be effectively 'captured'. Examples that fall into this category are the pre-eruption monitoring of large, active volcanoes; mapping ecotype boundaries in semi-arid marginal environments, as a contribution to the assessment of desertification; biomass, vegetation vigour and crop production estimates for homogeneous ecosystems and agricultural economies; locating desert locust breeding grounds; and monitoring changes of ocean water temperatures as a means of clarifying patterns and processes of upwelling. The challenge of the future is to use meteorological satellite data as a means of understanding relationships between climatic events and climatic change, on the one hand; and environmental responses eleswhere, e.g. in the oceans and large lake bodies and in ecosystem dynamics, on the other. The superficial 'housekeeping' role of these platforms is only one dimension of their real versatility.

In the sections that follow, basic details of the systems of meteorological data acquisition currently available are provided via the main agencies that provide archival and distribution facilities. All are sources of publications of various levels of sophistication.

1. <u>University of Dundee/Natural Environment Research Council</u>

Contact: University of Dundee Meteorological Station, Department of Electrical Engineering and Electronics, University of Dundee, Dundee, DD1 4HN (0382 23181, ext. 229 Telex 76293). The imagery comes from US polar orbiting satellites such as VHRR data (NOAA-5) and AVHRR (NOAA-7); a variety of formats are available:

1. For any day since 23 August 1976, black and white, visible and infrared images (9.5 x 6.25") from morning and evening

passes covering an area of 4700 x 3000km centred approximately on the British Isles and stretching from roughly Spitzbergen to North Africa and NE Atlantic to Eastern Europe. These are linearised to correct for Earth curvature distortion, which means that the scale is very nearly constant in all directions at all points on the image. Users can stipulate whether they do or do not require a latitude/longitude grid and land outline superimposed in white on each image. Archive data is supplied in the form of photofacsimile images.

Unless the user has a very clear idea of the cloud patterns in a given day's image, it is possible to request that Dundee select an example of, say, a depression or an anticyclone near or over the British Isles. The quality and resolution of the hard copy or prints is very good and the region covered means that there are normally many features of interest to discuss; obtaining images from consecutive days during a period of disturbed weather usually illustrates the mobility and changing nature of the synoptic cloud patterns. Matching visible and infrared images are instructive although the former are of limited use in winter because of poor illumination. Researchers may also obtain a computer compatible tape record for any part of the archive.

2. A browse file (8 x 10" contact printed sheets) containing 36 images per sheet from at least one channel (usually infrared) of every full pass recorded in the archive is also available. Each sheet represents a period of approximately a week and can be a useful guide to the selection of specific images discussed above which these files contain in reduced form. Synoptic cloud features, large fog banks, etc can be seen by the naked eye and more detail (eg snow and ice cover) by magnifying glass. The browse file is updated monthly, and new additions can be sent automatically to users requesting this service.

3. Sectionised electronic enlargements of prints discussed above (see 1) which are linearised but not gridded are available; the largest magnification is x 4.5 producing prints covering an area 275 x 275 km (10' x 10'). These images can be used to 'zoom in' on interesting sub-synoptic cloud features (eg lee waves) that may be noted on the smaller scale images to give an impression of the various degrees of scales of organisation of clouds. These images are time consuming and expensive to produce.

A free general information booklet outlining all meteorological satellite image products is available; it contains information on orbits, sensors and archives, and gives general advice on users' likely requirements and on unit costs for all types of products. Charges for NERC employed scientists and for NERC funded projects are less than for external users of this service.

4. Also marketed is a book of 50 meteorological satellite images, with captions, text and an appendix, suited to students, teachers and the general reader, viz; R R Fotheringham, The Earth's Atmosphere Viewed from Space, 1982, £4.60. A specialist two volume monograph, Cloud Cover Analysis of AVHHR Data, October 1978 - September 1982, by J M Anderson is published by the Carnegie Laboratory of Physics, University

of Dundee, at £2.00. This is an assessment of the density of cloud cover over sectors of the North Sea and the NE Atlantic in the vicinity of the Outer Hebrides.

2. European Space Agency

Contact: Meteosat Data Services (for images and data products), Meteosat Data Management Department, ESOC (European Space Operations Centre), Robert-Bosch-Strasse 5, D-6100 Darmstadt, Federal Republic of Germany.

Earth Observation Programme Office, ESA, 18 Avenue E Belin, 31055 Toulouse, Cedex, France. The imagery comes from the geostationary Meteosat 1 (images available from 23 November 1977 to 25 November 1979) and Meteosat 2 (images from August 1981 on), located over the intersection of the Equator and the Greenwich meridian.

1. Full disc Earth images centred on 0°E, 0°N for the visible, middle infrared (water vapour absorption) and thermal infrared bands, normally on a half-hourly basis (less frequent for the latter channel). These ungridded pictures are usually with 'disturbed' middle latitudes and a cloudy inter tropical zone; it is interesting to match these images with those from Dundee to compare perspectives and to give a different sense of scale. These are useful often to illustrate the extreme seasonal positions of the ITCZ-related cloud over Africa from, say, two images at the solstices (or August and February), as well as more general circulation patterns. A 20 x 20cm contact print costs less if ordered by a non-profit making user in the UK (costs escalate if a negative does not exist already in the archive). Transparencies may be purchased and can be used for making multiple copies of prints.

2. Enlargements of any full disc image are available either from a fixed grid of 8" x 8" boxes in which the user selects areas covered by one (visible band only), four or 16 elements or, for similar areas, selects a region by quoting the south-east corner's geographical coordinates. In addition there are two standard 'views' available - one of the British Isles, and the other of Europe. These images can be of value in looking in more detail at the cloud features, in the tropics and cost the same as images described in 1. They constitute the imagery products often used on the TV weather forecasts. North-West Europe and the British Isles are represented obliquely, with some loss of clarity compared to lower latitude areas.

3. Introduction to The Meteosat System, is a useful booklet of 12 pages which includes details of the satellite system ie radiometers, ground system, imagery processing, extraction of meteorological parameters, data collection and archiving. It also contains three full disc images and sectional images of Europe. The booklet is free of charge on request.

4. Meteosat System Guide (1980): this is a detailed manual on all aspects of the imaging system, archive and product specifications. It should be noted here that image analysis of Meteosat scenes can be undertaken at the NRSC.

5. Meteosat Data Services Newsletter: this is a useful (and free) guide to the range of products available and their prevailing costs. Some of the advertising features can make suitable posters for general teaching purposes.

6. Meteosat Image Bulletin: this monthly publication provides updating on the status of the archive and contains one full disc visible, infrared and water vapour image (at 12.00 GMT) for each day of the month in question. Available for inspection at the Meteorological Office, Bracknell, or directly on subscription.

7. Atlas of Meteosat Imagery: this contains 560 black and white plates of well selected and contrasting weather situations in a variety of latitudes. Chapters cover cloud classification, mesoscale cloud formations, synoptic scale cloud patterns, aspects of tropical cloud systems, and planetary scale cloud systems. All plates are accompanied by explanatory caption.

8. Researchers can acquire archive details on computer compatible tape (CCT) or high density tape (HDT). Each HDT has capacity for one day's images on all three channels; each CCT can contain only one image in the infrared water or visible channels (at half resolution).

NB: Although the NRSC, as part of the RAE Space Department, operates a ground station for the reception of the Meteosat data, this is for transmission to users such as the Meteorological Office. The NRSC is not an official archive, although it may be contacted to offer assistance and it does hold a limited number of scenes that are of particular interest to teachers. In addition to Meteosat, this reception, processing and distribution service extends to GOES and NOAA polar orbiting meteorological satellite data (see below).

3. National Oceanic and Atmospheric Administration

Contact: Satellite Data Services Division, NCC/NESDIS (National Climatic Center/National Environmental Satellite Data and Information Services), World Weather Building, Room 100, Washington D.C. 20233, USA. The imagery comes from US polar orbiting and geostationary satellites (Tiros, Nimbus, GOES) covering some regions complementary to those recorded at Dundee and Darmstadt. (It also includes Seasat sensor data.) HCMM data is separately archived (it is also available directly from the NRSC in the UK, as it was not a dedicated meteorological satellite).

1. Geostationary full disc visible and infrared images from 9 August 1976 from both GOES East (centred at 0°N, 75°W) and GOES West (centred at 0°N, 135°W) are available on a half-hourly basis as a 10" x 10" black and white contact print (plus a grid if required). These provide a different geographical context but are not normally of the same high quality as Meteosat images.

2. Sectorised scenes are available on a three hourly basis for infrared images throughout the period and with a similar frequency during daylight hours for visible. Each section

consists of data viewed for roughly 105° of latitude and 99° of longitude starting at 50° N and centred east/west at the subsatellite longitude.

3. A selection of 16-mm movie loops are available, made from half-hourly visible or infrared geostationary full disc images (80 or more frames) of mainly sectorised North American images of 1 to 2 km resolution (also available in enhanced infrared). Duplicates of 16-mm operational loops are used originally as tools for weather forecasters and provide a useful illustration of short term changes in weather features. An 'On-Line User Services' (OLUS) capability will be introduced by NESDIS in early 1986, with the objective of giving an on-line operational data base (both in near real time and retrospectively) to users of all types of US meteorological satellite data.

4. Stretched, gridded, pole-to-pole strips are available from the polar orbiters in both visible and infrared. These consist of three panels each 11.2 x 33.5 cm, which make up the complete strip. These are interesting in form but only of limited value because they are unrectified.

5. (a) National Environmental Satellite Service produce a catalogue of products, the latest edition of which is free; and the National Holdings of Environmental Satellites (also free) which are useful guides to all the material available, with examples and prices.
(b) Satellite Data Users Bulletin gives latest information on the status of data, together with developments in progress and future plans.

6. Environmental Satellites: Systems, Data Interpretation and Applications: the latest edition of this is a very useful and inexpensive introduction to the past, present and future of meteorological satellites, how they operate and the type of data produced, and how these are processed and disseminated. Techniques of image interpretation and how they are used practically is also discussed. This booklet contains 43 pictures which can be purchased with the booklet either as a black and white filmstrip or a set of slides (six, in colour). It should be noted that oceanographic as well as meteorological satellite data is included.

4. Other Resources

1. Collections of slides and films on meteorological satellite data and its applications are listed in Part 2.3.

2. A set of four large format black and white photographic prints of cloud patterns over the British Isles, NW Europe and the NE Atlantic are available at £1.00 each (postage included) from The Meteorological Office, Met. O.7a, London Road, Bracknell, Berks, RG12 2SZ. The set has been chosen to illustrate a variety of synoptic weather situations and uses visible and infrared spectral band images from NOAA 6 and 7, TIROS-N and Meteosat 2. A short explanatory caption is printed below each image frame.

3. Sat Pack 1 and 2 is a widely adopted teaching resource

giving the 'how and why' of weather satellites, and includes structured student exercises and teaching notes (see Part 2.5).

4. A recently published resource of particular interest to secondary schools and higher education establishments is: Weather, Radar and Satellite Images on the BBC Micro published by Computer Solutions for Science and Business, 73 Church Street, Malvern, Worcestershire, WR14 and developed and written by Dr A Eccleston. It is a package that contains 30 images illustrative of weather situations that frequently recur over the British Isles. It comprises two 80-track (or four 40-track) discs, accompanying software and a teaching manual. The latter incorporates background explanatory notes about the meteorological conditions and acquisition systems for each image. Student exercises are set out, and can be worked independently. They do, however, assume familiarity with basic meteorological theory. An interesting component of this package is the inclusion of animated sequences of ground-based radar and satellite imagery to forecast the growth and movement of rain areas over very short time periods (up to six hours). The product was made in close collaboration with the Meteorological Office; it costs £39.95.

5. NASA (Education Programs Branch) have published (1978) a useful booklet, Teacher's Guide for Building and Operating Weather Satellite Ground Stations for High School Science, by R J Summers and T Gotwald.

6. Feedback Instruments Ltd market a weather satellite data recording system ('WSR 513/515') that allows for colour enhancement and density slicing of imagery. Picture standard is 625/50 Hz giving video tape storage and CCTV display opportunities. An optional extra allows automatic operation from microcomputers. This facility may be of special interest to higher education establishments and to syndicates of schools or teacher centres. Full details from Feedback Instruments Ltd, Park Road, Crowborough, East Sussex, TN6 2QR (08926 3322).

PART 2.2: PUBLICATIONS

A. TEXTBOOKS

The rapid development of the technology of remote sensing, especially sensor capability and computerised image analysis procedures, has led to the inevitable obsolescence of texts that have not been subject to frequent revision. In common with any new scientific methodology, many general texts on remote sensing have appeared almost simultaneously and have strong similarities of structure, organization, and detailed content. The only distinguishing features often prove to be the case studies used, which usually reflect the specialism(s) of the author(s). The texts listed below are therefore selected from the full range of titles available, using the criteria of (i) up-to-date content, (ii) comprehensive and clearly-written coverage of the 'core' areas of the subject, or (iii) numerous and well-presented illustrations that complement the text. As remote sensing is such a visual medium, the last criterion is perhaps the most important of the three.

Titles that are preceded by an asterisk (*) are by British authors and contain case examples from the UK.

Texts for the general reader, or for anyone wishing to gain an essentially non-technical overview of the subject, are few. The only one currently available is:

Harper, D. (1983) (2nd Edition) Eye in the Sky: Introduction to Remote Sensing Multiscience Publications Limited, 1253 McGill College Street, Suite 175, Montreal, Quebec H3B 2Y5, Canada (C$18.50, plus $2.50 for postage outside Canada).

Another short, succinct text by Rudd, R. D. (1974) Remote Sensing: A Better View Duxbury Press is now in need of rewriting.

Although not a text in the strict sense, a short summary of remote sensing is provided by Bullard, R. K. and Larkin, P. J. (1981) First Steps in Remote Sensing Working Paper No. 3, North-East London Polytechnic, Department of Land Surveying, Dagenham, Essex (£3.50).

More substantial texts, aimed primarily at undergraduate and professional readership, are relatively numerous. Most would be suitable as a means of introduction and orientation for scientists and mathematicians without previous awareness of the aims, methods, materials and research potential of remote sensing. The titles given below are those which are commonly recommended on existing undergraduate and post-graduate courses in the U.K.

* Barrett, E. C. and Curtis, L. C. (1983) (New Edition) Introduction to Environmental Remote Sensing Chapman and Hall, London, Paperback, £14.45. Widely adopted as essential reading on many higher educational courses.

* Curran, P. J. (1985) Principles of Remote Sensing Longman, (£11.95, paperback). An attractive and well-ordered text, suitable for undergraduates and others with limited previous knowledge, which concentrates on the acquisition and interpretation of both aerial and space-derived imagery.

* Harris, R. (1986) Remote Sensing from Satellites Routledge and Kegan Paul, London, £6.95, Paperback.

Holz, R. K. (1984) (2nd Edition) The Surveillant Science: Remote Sensing of the Environment John Wiley, Chichester, £19.45, hardback. It is intended for the advanced student, especially post-graduates and higher education teachers, with emphasis on research. The structure of this text is based directly on the wavelength regions of the electromagnetic spectrum.

Lillesand, T. M. and Kiefer, R. W. (1979) Remote Sensing and Image Interpretation John Wiley, Chichester, £16.60. This text includes coverage of the elements of photogrammetry, as well as the more usual 'core' areas of remote sensing widely adopted in US colleges and universities.

Lintz, J. L. and Simonett, D. S. (Eds.) (1978) Remote Sensing of the Environment, Addison-Wesley, Reading, Massachusetts, US$32.50. A substantial and scholarly overview of the state of the art in the latter 1970s.

Richason, B. F. (1978) Introduction to Remote Sensing of the Environment Kendall/Hunt Publishing Company, Dubuque, Iowa 52001, US$27.50. Written specifically for the American undergraduate market, and supplemented by Laboratory Manual to Remote Sensing of the Environment $8.95. The latter contains a variety of image analysis exercises, all taken from landscapes of the U.S.A. Teachers at all levels would find several examples adaptable to areas and topics that have specific course or syllabus significance.

Sabins, F. F. (1978) Remote Sensing: Principles and Interpretation W. H. Freeman, Oxford, £11.80. Although in need of some updating, the core of this text is thoroughly relevant. There is a degree of bias in the examples towards earth science, but the illustrations are abundant.

Swain, P. H. and Davis, S. M. (Eds.) (1981) Remote Sensing: The Quantitative Approach McGraw-Hill, New York, $39.50. Well-integrated book that objectively examines the opportunities and limitations of data processing, image analysis and interpretation. An effective introduction to a potentially confusing and complex, but vital, area of remote sensing. Relevant to teachers, students, and personnel involved in commercial applications of data drawn from both remote sensing and other sources of information.

Although now surpassed by developments of technology and methods,there is still value in an effective overview edited by Estes, J. E. and Senger, L. W. (1974) entitled Remote Sensing: Techniques for Environmental Analysis and published by Hamilton, Santa Barbara, California (a subsidiary of John Wiley).

* The papers presented at a Royal Society Discussion Meeting in November 1982 by 24 researchers active in a wide variety of application fields, most of them British, was published in mid-1985 as The Study of the Ocean and the Land Surface from Satellites Houghton, J. T., Cook, A. H. and Charnock, H. (Eds.) Cambridge University Press, Cambridge, £35.00. Although now dated, a useful variety of examples and applications in the British context appears in:

* Barrett, E. C. and Curtis, L. F. (Eds.) (1975) Environmental Remote Sensing Vol. 1, Edward Arnold, London.

* Barrett, E. C. and Curtis, L. F. (Eds.) (1975) Environmental Remote Sensing Vol. 2, Edward Arnold, London.

* Peel, R. F. et al. (Eds.) (1977) Remote Sensing of the Terrestrial Environment Colston Papers, Vol. 16, Butterworths, London.

The fundamental reference text for remote sensing - the opus magnum of the subject, and unlikely to be surpassed - is:

Colwell, R. N. (Editor-in-Chief) and others (1983) (2nd Edition) Manual of Remote Sensing 2 volumes, American Society of Photogrammetry, US$125.00.

Vol. 1: Theory, Instruments and Techniques; Vol. 2: Interpretation and Applications. A total of 2,724pp., including 280 colour plates of imagery, a comprehensive index and a full glossary of technical terms. A thorough, well-edited and authoritative work that is remarkably good value. No serious student or institution with a commitment to remote sensing can afford to be without this work. Although it is not intended as a text, individual chapters or sequences of chapters will serve this purpose well. The substantive content is, perhaps inevitably, biased towards North American practice and experience. The major disadvantage of this encyclopaedic tome is the difficulty of maintaining it as an up to date survey of the state of the art of remote sensing. There are no publicised plans to issue occasional supplements. A good deal of the content on the physical basis of the subject is standard, of course. For this reason, copies of the first edition (1975) are worthy of acquisition.

The Manual may be obtained from the American Society of Photogrammetry, 210 Little Falls Street, Falls Church, Virginia 22046. Lower prices apply to members of the Society and to academic libraries worldwide. A detailed publicity brochure is available, on request; details of the Society's other relevant publications might also be requested. The most generally useful of these is the Multilingual Dictionary of Remote Sensing and Photogrammetry (1984), with over 1,700 terms defined in English and translated into French, German, Russian, Italian, Spanish, and Portuguese. Paul, S. (1982) Airborne-Spaceborne Remote Sensing Dictionary Laffont, Paris is a further addition to the terminological literature.

There are numerous specialised texts on specific aspects of remote sensing, particularly those concerned with research and management of environmental resources in defined areas. From a potentially long list, the following are selected:

1. General

Clough, D. J. and Morley, L. W. (Eds.) (1977) Earth Observation Systems for Resource Management and Environmental Control Plenum Press, New York.

* GRAMPIAN REGIONAL COUNCIL (1983) Remote Sensing and Resource Planning in Scotland Grampian Regional Council, Aberdeen.

Johannsen, C. and Sanders, J. (Eds.) (1981) Remote Sensing for Resource Management Soil Conservation Society of America, Washington D.C.

Schanda, E. (Ed.) (1976) Remote Sensing for Environmental Sciences Springer, Heidelberg, FRG.

Williams, R. S. and Carter, W. D. (Eds.) (1976) ERTS-1: A New Window on Our Planet U.S. Geological Survey Professional Paper 929: USGPO, Washington D.C.

2. Physical Basis and Spectral Regions

Anon. (1975) Remote Sensing Energy-Related Studies Hemisphere Publishing Co., Washington D.C.

Dellwig, L. F. et al. (1972) Radar Remote Sensing for Geoscientists University of Kansas Center for Research, Remote Sensing Laboratory, Lawrence, Kansas, USA.

Egan, W. G. (1985) Photometry and Polarization in Remote Sensing Elsevier, New York.

Kovaly, J. (Ed.) (1980) Synthetic Aperture Radar Artech House, New York.

Long, M. W. (1975) Radar Reflectivity of Land and Sea Lexington, Boston, Massachusetts.

Maby, F. T., Moore, R. K. and Fung, A. D. (1981-83) Microwave Remote Sensing: Active and Passive 2 volumes, Addison-Wesley, New York.

Slater, P. N. (1980) Remote Sensing: Optics and Optical Systems Addison-Wesley, New York.

3. Image Processing, Analysis, and Interpretation

Andrews, H. C. and Hunt, B. R. (1977) Digital Image Restoration Prentice-Hall, Englewood Cliffs, New Jersey.

Anon. (1981) Computer Mapping of Natural Resources and the Environment: Applications of Satellite-Derived Data Vol. 10, Harvard Library of Computer Graphics and Mapping series (this revises vol. 4 of the same series, with the same title), HLCG (850 Boylston Street, Chestnut Hill, Massachusetts 02167,

U.S.A.) and Addison-Wesley, New York.

Boerner, W. M. (Ed.) (1984) Inverse Methods in Electromagnetic Imaging D. Reidel, Dordrecht.

Castleman, K. R. (1979) Digital Image Processing Prentice-Hall, Englewood Cliffs, New Jersey.

Devjver, A. and Kittler, J. (1983) Pattern Recognition: A Statistical Approach Prentice-Hall, Englewood Cliffs, New Jersey.

Gonzalez, R. C. and Wintz, P. (1977) Digital Image Processing Addison-Wesley, New York.

Haralick, R. M. and Simonett, D. S. (1983) Image Processing for Remote Sensing Addison-Wesley, New York.

* Hood, R. M. (1982) Digital Image Processing of Remotely-Sensed Data Academic Press, London and New York.

Hopkins, P. F. (Ed.) (1984) Extraction of Information from Remotely-Sensed Images American Society of Photogrammetry.

Schowengerdt, R. A. (1983) Techniques for Image Processing and Classification in Remote Sensing Academic Press, New York and London.

Tou, J. T. and Gonzalez, R. C. (1979) (3rd Edition) Pattern Recognition Principles Addison-Wesley, New York.

4. Applications in Atmospheric Science

* Anon. (1980) Remote Sensing of Snow and Ice UNESCO, Paris.

* Barrett, E. C. (1974) Climatology from Satellites Methuen, London.

* Barrett, E. C. and Martin, D. W. (1981) The Use of Satellite Data in Rainfall Monitoring Academic Press, London and New York.

* Fotheringham, R. R. (1982) The Earth's Atmosphere Viewed from Space University of Dundee, Department of Electrical Engineering and Electronics.

* Henderson-Sellers, A. (Ed.) (1984) Satellite Sensing of a Cloudy Atmosphere Taylor and Francis, London.

* Houghton, J. T. et al. (1984) Remote Sounding of Atmospheres Cambridge University Press, Cambridge.

See also Cracknell, A. P. (Ed.) (1981) under Marine and Oceanographic Applications.

5. Applications in Hydrology and Water Resources

Anon. (1980) Remote Sensing of Snow and Ice UNESCO, Paris.

Fraysse, G. (Ed.) (1980) *Remote Sensing Applications in Agriculture and Hydrology* A. A. Balkema, The Hague.

Salomonson, V. V. and Bhavsar, P. D. (Eds.) (1980) *The Contribution of Space Observation to Water Resource Management* Pergamon Press, Oxford and New York.

See also Cracknell, A. P. (Ed.) (1981) under *Marine and Oceanographic Applications*.

6. Marine and Oceanographic Applications

* Allan, T. D. (Ed.) (1984) *Satellite Microwave Remote Sensing* Ellis Horwood, Chichester.

Anon. (1980) *Remote Sensing of The Oceans* UNESCO, Paris.

* Cracknell, A. P. (Ed.) (1981) *Remote Sensing in Meteorology, Oceanography and Hydrology* Ellis Horwood, Chichester.

* Cracknell, A. P. (Ed.) (1983) *Remote Sensing Applications in Marine Science and Technology* D. Reidel, Dordrecht.

Gautier, C. and Fieux, M. (Eds.) (1984) *Large-Scale Oceanographic Experiments and Satellites* D. Reidel, Dordrecht.

Gierloff-Emden, H. G. (1977) *Orbital Remote Sensing of Coast and Off-Shore Environments: A Manual of Interpretation* Walter de Gruyter.

Gower, J. F. R. (Ed.) (1981) *Oceanography From Space* Plenum Press, New York.

Leibert, A. (1982) *The Benefit of Environmental Satellites to Offshore Industries* Online Publications, Northwood, Middlesex.

Massin, J. M. (Ed.) (1984) *Remote Sensing for the Control of Marine Pollution* Plenum Press, Oxford.

Maul, G. A. (1985) *Introduction to Satellite Oceanography* Martinus Nijhoff, The Netherlands.

Nihoul, J. C. J. (Ed.) (1984) *Remote Sensing of Shelf Sea Hydrodynamics* Elsevier, Amsterdam.

* Robinson, I. S. (1985) *Satellite Oceanography* Ellis Horwood, Chichester.

7. Applications in Geology, Geomorphology, and Soil Science; Civil Engineering

Carter, W. D. (Ed.) (1980) *Remote Sensing and Mineral Exploration* Pergamon Press, Oxford.

Kennie, T. J. M. and Matthews, M. C. (1985) *Remote Sensing in Civil Engineering* Surrey University Press, Guildford.

Siegal, R. S. and Gillespie, A. R. (1980) Remote Sensing in Geology John Wiley, Chichester.

Smith, W. L. (Ed.) (1977) Remote Sensing Applications for Mineral Exploration Dowden, Hutchinson and Ross, Pittsburgh, USA.

Townsend, J. R. G. (Ed.) (1981) Terrain Analysis and Remote Sensing Allen and Unwin, London.

Verstappen, H. Th. (1977) Remote Sensing in Geomorphology Elsevier, Amsterdam.

Watson, K. and Regan, R. D. (Eds.) (1983) Remote Sensing Society of Exploration Geophysicists, Tulsa, Oklahoma.

8. Other Texts

Berg, A. (1980) Applications of Remote Sensing to Agricultural Production Forecasting A. A. Balkema, The Netherlands.

Ebert, J. I. (1985) Remote Sensing in Archaeology and Cultural Resources Management Martinus Nijhoff, The Netherlands.

El-Baz, F. (Ed.) (1984) Deserts and Arid Lands: Remote Sensing Martinus Nijhoff, The Netherlands.

* Fuller, R. M. (Ed.) (1983) Ecological Mapping from Ground, Air and Space Institute of Terrestrial Ecology (NERC), Cambridge.

Fraysse, G. (Ed.) (1980) Remote Sensing Applications in Agriculture and Hydrology A. A. Balkema, The Netherlands.

Lindgren, D. T. (1985) Land-Use Planning and Remote Sensing Martinus Nijhoff, The Netherlands.

Lund, T. (Ed.) (1979) Surveillance of Environmental Pollutants and Resources by Electromagnetic Waves D. Reidel, Dordrecht.

Woodwell, G. M. (Ed.) (1984) The Role of Terrestrial Vegetation in the Global Carbon Cycle: Measurement by Remote Sensing John Wiley, Chichester.

All of the above titles published by Martinus Nijhoff (The Hague and Lancaster, U.K.) are part of a series titled Remote Sensing of Earth Resources and Environment. Several additional texts are in advanced preparation, and should be published by early 1986. These include Coastal Resources Management; Petroleum Resources, and Geobotanical Prospecting.

B. AERIAL PHOTOGRAPHY and PHOTOGRAMMETRY

The much longer-established methods of acquisition and interpretation of vertical air photos (black and white; true colour and infrared false colour), and the science of photogrammetry, has produced a long lineage of textbooks. The following titles are either modern surveys of the field or those which have achieved classic status. An asterisk (*) indicates British authorship, or a substantially British context.

1. General

Anon. (1960) Manual of Photographic Interpretation American Society of Photogrammetry.

Anon. (1968) Manual of Colour Aerial Photography American Society of Photogrammetry.

Anon. (1976) Survey Manual prepared jointly by staff of Division of Planning Intelligence, Department of the Environment and Fairey Surveys (now Clyde Surveys) describing land-use classification, image interpretation techniques, production methods, etc., used for the DoE-commissioned National Land-Use Survey of the Developed Areas of England and Wales by Remote Sensing, which used aerial photography and Landsat imagery (subsequently analysed and mapped by automated cartography), Clyde Surveys.

Anon. (1980) (4th Edition) Manual of Photogrammetry American Society of Photogrammetry, 2 volumes.

Avery, T. E. (1977) (3rd Edition) Interpretation of Aerial Photographs Burgess Publishing Company, Minneapolis, Minnesota.

Brock, G. C. (1967) The Physical Aspects of Aerial Photography Dover Publications, New York.

* Burnside, C. (1979) Mapping from Aerial Photographs Crosby, Lockwood and Staples, London.

* Curran, P. J. (1981) Remote Sensing: The Role of Small Format Light Aircraft Photography Geographical Papers, No 75, Department of Geography, University of Reading.

* Dickinson, G. C. (1981) (2nd Edition) Maps and Air Photographs Edward Arnold, London.

ITC Textbooks in Photogrammetry; Air Photo Interpretation; Aerial Survey. Available in various languages and as separate chapters, which are periodically revised. ITC, Enschede, The Netherlands.

* Kilford, W. K. (1980) (4th Edition) Elementary Air Survey Pitman, London.

Leuder, D. R. (1959) Aerial Photography Interpretation McGraw-Hill, New York.

* Lo, C. P. (1976) Geographical Applications of Aerial Photography David and Charles, Newton Abbot.

Moffitt, F. H. and Mikhail, M. (1981) Photogrammetry International Textbook Company, Scranton, Pennsylvania.

Mollard, J. D. (1960) Air Photo Analysis and Interpretation Bellhaven House Limited, Scarborough, Ontario, Canada.

Nagao, M. and Matsuyama, T. (1980) A Structural Analysis of Complex Aerial Photographs Plenum Press, New York.

Spurr, S. H. (1960) Photogrammetry and Photointerpretation Ronald Press, New York.

Strandberg, C. H. (1967) Air Discovery Manual John Wiley, Chichester.

Sully, G. B. (1970) Aerial Photo Interpretation Bellhaven House Limited, Scarborough, Ontario, Canada.

Wolf, P. R. (1974) Elements of Photogrammetry McGraw-Hill, New York (with an accompanying Solutions manual for teachers and students).

Geographical Interpretation of Aerial Photographs Bundesanstalt für Landeskunde und Raumforschung (1952-1974), Bad Godesberg, FRG.

2. Topical and other Specialist Texts

Allum, J. A. E. (1966) Photogeology and Regional Mapping Pergamon Press, Oxford and New York.

Anon. (1966) Aerial Photo-Interpretation in Classifying and Mapping Soils Agriculture Handbook 294, U.S. Department of Agriculture, USGPO, Washington D.C.

Anon. (1968) Aerial Surveys and Integrated Studies. Proceedings of the Toulouse International Conference UNESCO, Paris.

Anon. (1969) Forester's Guide to Aerial Photo-Interpretation Agriculture Handbook 308, U.S. Department of Agriculture, USGPO, Washington D.C.

Anon. (1972) The Use of Aerial Photography in Countryside Research Report CCP 54, Countryside Commission, London.

Branch, M. C. (1971) City Planning and Aerial Information Harvard University Press, Cambridge, Massachusetts.

Campbell, J. B. (1984) Mapping the Land: Aerial Photography for Land Use Information Association of American Geographers, Washington DC.

* Carroll, D. et al. (1978) Air Photo Interpretation for Soil Mapping Soil Survey of England and Wales, Monograph No 5.

Devel, L. (1969) Flights into Yesterday - The Story of Aerial Archaeology St Martin's Press, New York.

Edwards, G. J. and Blazquez C. H. (Eds) (1984) Color Aerial Photography in the Plant Sciences American Society of Photogrammetry, Falls Church, Virginia.

Goodier, R. (Ed.) (1971) The Application of Aerial Photography to the Work of the Nature Conservancy The Nature Conservancy Council, Edinburgh.

Hammond, R. (1967) Air Survey in Economic Development Elsevier, Amsterdam.

Howard, J. A. (1970) Aerial Photo-Ecology Faber and Faber, London.

Lattman, L. H. and Ray, R. G. (1965) Aerial Photographs in Field Geology Holt, Rinehart and Winston, New York.

Lyons, T. R. and Avery, T. E. (1977) Remote Sensing: A Handbook for Archaeologists and Cultural Resource Managers U.S. Department of the Interior, National Park Service, Washington D.C.

Miller, V. C. (1961) Photogeology McGraw-Hill, New York.

Paine, D. P. (1981) Aerial Photography and Image Interpretation for Resource Management John Wiley, Chichester (also widely-used as a general text).

Ray, R. G. (1960) Aerial Photographs in Geological Interpretation and Mapping US Geological Survey, Professional Paper 373, USGPO, Washington D.C.

Way, D. (1968) Air Photo Interpretation for Land Planning Harvard University Press, Cambridge, Massachusetts.

Way, D. (1973) Terrain Analysis: A Guide to Site Selection Using Aerial Photographic Interpretation Dowden, Hutchinson and Ross, Pittsburgh, USA.

* White, L. P. (1977) Aerial Survey and Remote Sensing for Soil Survey Clarendon Press, Oxford.

* Wilson, D. R. (Ed.) (1980) Aerial Reconnaissance for Archaeology London Council for British Archaeology, Research Report No 12.

Eastman Kodak (Rochester, New York 14650, U.S.A.) have published several brochures and papers on film types and properties as well as more general attributes of aerial photography. They are listed and summarised in the company's Index to Kodak Information, obtainable on request. Titles include Photography from Light Planes and Helicopters; Photo-Interpretation for Land Managers; Photointerpretation for Planners and Aerial Photography as a Planning Tool. Single copies of each publication are supplied free of charge. Stocks are not held by the U.K. offices of Kodak.

There are a limited number of manuals and atlas-style publications devoted to air photo stereo pairs, which may be viewed three-dimensionally given the availability of suitable equipment. In all cases, there is accompanying text giving locational detail and offering help with detailed interpretation. Among these publications may be listed:

1. Anon. (1968) Stereo Atlas Hubbard Press, Northbrook, Illinois (available in U.K. from Hestair Hope). Fifty black and white stereograms selected from the extensive collection of vertical air photos held by the U.S. Geological Survey. Most are reproduced facing a map extract of the same location, and are accompanied by an oblique aerial photographic view. The emphasis is on geological and geomorphological features; background detail is economical, but the examples all lend themselves to straightforward interpretation exercises by relatively inexperienced students, from 15+. Teachers, however, are required to devise their own questions.

2. Wanless, H. R. (1973) (2nd Edition) Aerial Stereo Photographs Hubbard Press, Northbrook, Illinois (available in U.K. from Hestair Hope). Ninety-two plates of vertical air photo stereo pairs, presented in a ring binding. The emphasis is, again, on natural landscape features, especially geomorphological patterns and processes. The examples are all from the coterminous United States. Each photo pair is accompanied by regional description, matching USGS Quadrangle topographic sheet, and details of scale, date, etc.

3. There is an accompanying booklet of 48pp. by Bowyer, R. E., and Snyder, P. B. Aerial Stereo Studies, containing eight graded interpretational questions on each of the 92 images reproduced in Wanless. These two complementary publications provide a sound basis for consolidating course work in earth science and physical geography through the medium of aerial photography. They equally provide a useful introduction to the essentially qualitative procedures of interpretation for inexperienced students from 15-19 (including first-year undergraduates). The exercises can be used 'ready-made' by teachers, as the examples are models of their kind, most of them developed at a spatial scale that is well in excess of any equivalent British examples. The suggested exercises vary from single line questions to essay assignments calling for further reading and air photo study of similar landscape forms.

All three publications are available from Hestair Hope Ltd (St Philip's Drive, Royton, Durham OL2 6AG) at the very low price of £1.80 each and are therefore especially attractive to schools and colleges wishing to build up multiple copies of teaching collections of air photos.

4. Anon. (1968) Stereo Atlas American Geological Institute. This is a more sophisticated publication, with much the same format as the publications noted above. Coverage and content is strictly geological.

5. Anon. Relief Form Atlas, 1985, £57.00, Nathan, Paris. A Relief Form Atlas was first published by the Institut Géographique National (France) in 1956. It met with great success and a new edition was necessary. This is now available in a greatly improved and extended version, using new documents, taking into account new developments in photography, mapping and remote sensing. A glossary of the different geological and geomorphological terms used in the book is also included.

6. Anon. (1970-1971) An Airphoto Atlas of the Rural United States Agriculture Handbooks Nos 372; 384; 406; 409 and 419, U.S. Department of Agriculture, USGPO, Washington D.C.

7. Garver, J. B. et al. (1984) (3rd Edition) Atlas of Landforms John Wiley, Chichester, £39.50.

Additional related publications containing student interpretation exercises based on published stereo pairs and single photographs are listed in Part 2.5.

3. Oblique Aerial Photography

Although oblique photos lack quantitative attributes, they nonetheless provide a visual perspective that can be of value in a wide variety of contexts. Obliques are very commonly used as illustrations in textbooks and in public examinations. Innovative teachers in junior and middle schools have found them to be of value in introducing or reinforcing basic concepts of environmental awareness. In the fields of archaeology, historical geography, and socio-economic history, oblique air photos have been extensively used to identify the evidence of settlement and other features that have obscurely visible ground-level forms.

From a very long list of publications that use oblique aerial photographs, wholly or in part, the following few are selected. All are confined to the British Isles.

Booth, S. (1979) Aerofilms Book of England from the Air Blandford, £5.75 (Available from Aerofilms Ltd).

Stonehouse, B. (1982) Aerofilms Book of Britain from the Air Weidenfield & Nicolson, £18.95.

Johnson, P. (1984) Aerofilms Book of London from the Air Weidenfield & Nicolson, £12.95.

Campbell, J. (1984) Aerofilms Book of Scotland from the Air Weidenfield & Nicolson, £12.95.

Kiely, B. (1985) Aerofilms Book of Ireland from the Air Weidenfield & Nicolson, £12.95.

Frere, S. S. and St Joseph, J. K. (1983) Roman Britain from the Air Cambridge University Press, Cambridge.

Hudson, K. (1985) Industrial History from the Air Cambridge University Press, Cambridge.

St Joseph, J. K. (1979) (2nd Edition) Uses of Aerial Photography John Baker, London (some of the examples are from overseas).

Walker, F. (1953, with revisions) Geography from the Air Methuen, London.

C. ATLASES AND PICTURE BOOKS OF IMAGERY

These large format books may be variously regarded as coffee table publications, absorbing records of imagery in their own right or as basic source-books for private study and class exercises. At least one of them justifies a place in most school and public libraries. Teachers and lecturers may be able to use selected plates as a basis for making class copies, once copyright restrictions have been cleared with publishers.

The titles listed here are limited to satellite and spacecraft-derived images of the Earth. In all cases, captions provide a description of the major landscape features revealed on each image scene. An introduction providing an overview of remote sensing and an account of how the images are obtained is also included in each book.

Early Gemini and Apollo true colour oblique photographs are reproduced in a number of books, all of which are out of print. These include: Lowman, P. D. (1968) Space Panorama Weltflugbild, Zurich (Switzerland). Nicks, O. W. (Ed) (1970) This Island Earth NASA (USGPO), 182pp (Pub SP-250), and Earth Photographs From Gemini III, IV and V NASA (USGPO), 266pp (Pub SP-129) 1967. It was these photographs that first captured the attention and the imagination of the world's public. They retain all of their original visual appeal and continue to be used as illustrations in textbooks (and by advertising agencies!). Examples of both land surfaces and of weather patterns in coastal and oceanic areas are included in roughly equal measure.

A compilation of many of the best photographs from manned space missions appears in Bodechtel, J. and Gierloff-Emden, H. G. (1974) The Earth From Space, David and Charles, 176pp, translated from an earlier German edition.

A good selection of both oblique and vertical imagery, in infrared as well as visible wavelengths, from the US Skylab mission, 1973-74, is reproduced in Skylab Explores the Earth NASA, (USGPO), 1975. In this case, there are more substantial chapters and sections providing details of the remote sensing experiments conducted from Skylab and providing detailed interpretation of image details.

A combination of Skylab and earlier photography of the earth (most of it confined to 30° north and south of the Equator) is integrated in Dickson, P. (1977) Out Of This World Dell Publishing Co, USA.

The first atlas of imagery from the Landsat programme was published by NASA in 1976 as Mission to Earth USGPO. This substantial volume includes over 400 plates, nearly all of which are colour composite images obtained by Landsat 1 between 1972 and 1974. Coverage is worldwide, though there is a bias toward North America. In default of a compendium of more recent Landsat images, this collection remains the most comprehensive yet published; at $32.00 it is remarkable value for money. Each image is described in a caption of about 200 words, where the emphasis of the authors tends towards geological and geomorphological detail.

In 1978, NASA published an accompanying booklet, Educator's Guide to Mission to Earth, containing ideas on lesson planning and class exercises using the original publication. The age level appropriate to most of these suggestions is 9-15.

An atlas of Landsat colour composite mosaics of East and West Germany, Austria and Switzerland was published by Georg Westermann in 1978. Although a pioneer attempt at the compilation of a national atlas using Landsat as the primary base, the quality of both feature enhancement and of image reproduction is variable. The full title is: Bodechtel, J. et al, (Eds) Weltraumbild Atlas (Deutschland, Osterreich, Schweiz). It is available from Georg Westermann Verlag, PO Box 3320, Georg Westermann Allee 66, D-3300 Braunshweig, West Germany. In 1984, Kummerly and Frey produced a Landsat Image Atlas of Switzerland. The Science Press, Beijing, published an encylopaedia in 1984 Landsat Atlas of China containing more than 1200 plates. It is essentially a series of image maps, and is described in detail on p. 18.

An altogether more ambitious and innovative atlas from Westermann utilising Landsat images at various scales, is Diercke Weltraumbild Atlas (1981). This book was published in 1984 in an English edition, as Images of the World: An Atlas of Satellite Imagery and Maps by Collins/Longman (available from the publishers at Longman House, Burnt Mill, Harlow, Essex CM20 2JE, at £12.95). It contains 124 satellite images and 150 corresponding maps, taken from locations all over the world. The important innovation is that images have been given a normal colour coding, on the assumption that false colours present interpretational problems for students. The images and maps are complementary, the annotation and symbolism of the latter assisting with the full appreciation of the former; however, map and image correspondence is not always as close as might be expected. The content of the atlas is arranged into eight thematic sections, which makes it particularly relevant to geography at secondary level. An introductory section presents the theory and application potential of remote sensing in general and there is a concluding chapter by R. M. Smith that makes suggestions for the use of the atlas in day-to-day geography teaching. Most of the images used are from Landsat 1 and 2, so recent improvements in spatial resolution are not represented.

Two books by the same author, Charles Sheffield, are devoted to Landsat colour composite images distinguished by very high standards of presentation. Most of the examples were specially commissioned and are the product of image enhancement techniques carried out by Earth Satellite Corporation (Washington, D.C.). No other publications achieve the excellence of reproduction of original imagery, which is justification enough for their availability in both general and specialist libraries. The accompanying text is more appropriate to the interested layman or inexperienced student, giving both books a wide appeal. The first, Earthwatch: A Survey of the Earth from Space (1981), 80 plates and 60 explanatory sketch maps, is largely devoted to recording the world's natural environment and costs £12.95. The second, Man on Earth: The Marks of Man: A Survey from Space (1983), 70 plates and 90 maps is specific to the impact of human societies and man's role in changing the face of the Earth; it costs £12.95. Coverage in both cases is global. The publishers of both are Sidgwick and Jackson, 1 Tavistock Chambers, Bloomsbury Way, London WC1A 2SG.

A recently published selection of Landsat colour images, integrated with natural colour photographs taken from the Space Shuttle and earlier manned Earth orbital missions, gives Francis, P. and Jones, P. Images of Earth a distinctive flavour. The overall theme is that of the historical development of landscape in different parts of the world, but scenes that are largely free of visible human impact are also included. This is a visually arresting presentation, with the accompanying maps and text providing an essentially non-technical interpretation of the chosen images. Published in December 1984 by George Philip, 12-14 Long Acre, London WC2 9LP; it costs £12.95. The 160 pages include 80 plates, which are reproduced to a standard that does credit to the art of the photolithographer. A special feature is the inclusion of several Landsat thematic mapper images, chosen to illustrate places in the international news during the first half of the present decade.

Increasingly, revised editions of well-established general and school atlases, as well as new publications, include selections of satellite imagery (usually Landsat) to add variety, contrast and interpretational interest. They often serve to emphasize the conventions of atlas cartography. An example is the latest edition (1984) of Philips Certificate Atlas.

An atlas designed specifically for use in Canadian schools, but worthy of wider adoption is Kirman, J. A Primer for Satellite Maps. Professor Kirman has been responsible for the development of a research programme at the Faculty of Educational Studies, University of Alberta, Calgary, devoted to the interpretational ability of students, between 11 and 18 years, applied to both black and white and false colour Landsat images. This atlas is partly a product of this experience, and partly a basic teaching resource for Canadian school teachers. Published in 1978, at C$40.00 by Puckrin's Production House, 35 Mill Drive, St Albert, Alberta, Canada T8N 1JS.

The Atlas for the Interpretation of Multispectral Aerospace Photographs: Methods and Results, published jointly by the Soviet Academy of Science, Moscow and the Academy of Science of the German Democratic Republic (East Berlin) in 1982 is a large format and high quality collection of images at a wide variety of scales. It is particularly valuable as a unique portfolio of imagery products from Soviet and East European remote sensing programmes, which are otherwise not available in the West. The English language edition provides descriptive captions of salient scene features and actual or potential application programmes. There are 83 plates, with the complete portfolio costing 480 East German Marks.

Published in June 1985 is Bullard, R. K. and Dixon-Gough, R. W. Britain from Space: An Atlas of Landsat Images Taylor and Francis, Basingstoke (£12.50). This atlas contains full coverage of the UK using colour Landsat scenes, with each accompanied by an annotated map and descriptive legend. Examples of processed images, to illustrate how thematic information can be displayed, are included and there is also a selection of other satellite sensor imagery of the country. The various proven and potential applications of satellite data to land cover mapping are acknowledged by reference to selected images. A basic introduction to concepts and techniques of analysis in remote sensing; a glossary of terms and a directory of research centres and equipment manufacturers complete the content. A set of 35 mm slides of selected images and a Workbook of student exercises are planned as ancillary materials.

One problem that afflicts all atlas-style publications is the length of time, at least 3-4 years, required for all phases of production. The imagery is therefore out of date in the sense that the spectral, radiometric and spatial resolutions of imaging systems has subsequently improved, justifying almost immediate substantial revision!

D. THE TECHNICAL LITERATURE OF REMOTE SENSING

Monographs, reports, memoranda, conference and workshop proceedings, and refereed papers in scientific journals have proliferated in the last 10-15 years. Because of the rapidity of technological change, many publications rapidly lose currency; this is especially true of the literature concerned with sensor types, platform and data analysis techniques. Retrospective bibliographical searches can often be safely confined to the last five or six years if the objective is to gain an overview of the state of the art. Specialists in particular application fields, e.g. structural geology, agricultural crop monitoring, coastal sediment dynamics, or intra-urban traffic control (to name a few) may derive more value from a bibliographical search over a longer period of time. Papers published 12-15 years ago may still retain some intrinsic value by virtue of case study area or the quality of interpretational results.

The dedicated researcher wishing to keep fully informed and up to date must have access to published bibliographies and abstracting services. The libraries of most educational institutions, as well as certain public libraries, offer computerised on-line citation searches that will identify relevant titles. However the success of such an approach depends on the quality of the database and the skill of the user in selecting key words. The most useful sources of bibliographical information are:

Anon. (1983) Imagery and Remote Sensing Abstracts Vol. 1. Five issues p.a. covering approximately 1,000 abstract entries. Apart from the conventional areas of the subject, aspects of military applications, deep space imaging missions, and satellite tracking are also included. Britanavia Publications, London.

Geo Abstracts (Part G, Cartography and Remote Sensing). Six issues annually, published since 1974. This is one of several separate volumes that cover the entire field of geography, oceanographic and atmospheric sciences, hydrology, and ecology. Geo Abstracts, Norwich NOR 8BC (University of East Anglia).

Technology Applications Center, University of New Mexico, Albuquerque, USA. Bibliographic compilation on specific topics is produced based on Reports from numerous US and other foreign agencies, e.g. NOAA; NASA; NESDIS; RESENA; RESORS (Canada); Earth Satellite Corporation, JPL, TAC; ERIM: IBM (Federal Systems Division) and National Technology Information Service (NTIS). The latter agency is responsible for regular updating of continuing NASA and NOAA bibliographies. All NTIS reports are indexed, and available from DRIC, Station House Square, St Mary Cray, Orpington, Kent BR5 3RE. (See also p. 48.)

For older sources, see also Krumpe, P. F. (1976) The World Remote Sensing Bibliographic Index, Tensor Industries, USA.

Quarterly Review of Literature, Technology Applications Center, University of New Mexico, Albuquerque, New Mexico 87131. Published since 1975, with a strong bias towards American sources. Annual subscription is US$85.00. (TAC also provides catalogues of Landsat, Gemini, and Apollo mission imagery.)

RESENA (Remote Sensing and Nature) bibliographies on remote sensing, starting with Technical Report No. 1 (1980) and followed by several other consolidated literature citations on specific topics. Details from RESENA Group, 1530 Wolf Run, College Station, Texas 88740, USA.

Vance Bibliographies The Public Administration Series of bibliographic indexes has been published for over 15 years, and incorporates a wide range of topics in social science, architecture, planning, and political administration. Several titles are specific to applications of remote sensing, e.g. Report No. P1460 Aerial Photography Monographs (1984) 25pp. Lists of titles in print are available from Vance Bibliographies, PO Box 229, Monticello, Illinois 61856, USA.

Bibliography of Remote Sensing Publications, University of Kansas Center for Research (1981). A numerical listing with abstracts of all reports published by the Remote Sensing Laboratory between 1964 and 1980 and of all published journal papers by RSL staff during the same period. 724 entries, 184pp., US$17.00 (overseas). A supplement for 1981-1983 was published in 1985 (60pp., US$16.00). This center has been particularly strong on research into microwave remote sensing. University of Kansas Center for Research, Remote Sensing Laboratory, 2291 Irving Hill Drive, Campus West, Lawrence, Kansas 66045, USA.

International Aerospace Abstracts (IAA) confined to conference papers, journal articles, books, journal translations, and published monographs; remote sensing is only one component of an extensive field thus covered. It is complementary in many ways to the bi-monthly STAR, published by NASA. Published by the American Institute of Aeronautics and Astronautics, Technical Information Service, 555 West 57th Street, 12th

Floor, New York, New York 10019, USA.

NASA: Remote Sensing of Earth Resources A literature survey with indexes. First published in 1970, and periodically up-dated as Earth Resources: A Continuing Bibliography NASA, Washington DC.

Science Reference Library (Holborn) division of the British Library, 25 Southampton Buildings, London WC2, 01-405 8721, offers a free bibliographic search service for short specific enquiries; on-line searches are also undertaken using computer databases such as Geo Archives, usually by appointment (and at cost).

The major journals dedicated to the publication of remote sensing articles are listed below. All of them are multi-disciplinary and international in outlook, so that papers specific to research carried out within the British Isles or by British scientists overseas will be widely-scattered.

1. International Journal of Remote Sensing (Vol. 1, 1980). 12 issues p.a.; £145 annually, £49 to members of the Remote Sensing Society. Subscribers also receive Remote Sensing Society's News and Letters, issued quarterly, consisting of short contributions of immediate currency. The journal is an official publication of the Remote Sensing Society, has a British editorial board, and contains a balance of original contributions and commissioned review papers. The publishers are Taylor and Francis, Rankine Road, Basingstoke, Hampshire RG24 0PR from whom sample copies are available.

2. Remote Sensing of Environment (Vol. 1, 1970). Mostly original research papers, each issue is devoted to a specific technique, theoretical topic, or application example. The editorial board and the large majority of articles are from North America. Six issues p.a. Subscription US$139.00. Published by Elsevier, New York and Amsterdam. Sample copies from Elsevier Science Publishers, PO Box 211, 1000 AE Amsterdam, The Netherlands.

3. Photogrammetric Engineering and Remote Sensing (previously Photogrammetric Engineering) (Vol. 1, 1934). The official journal of the American Society of Photogrammetry and Remote Sensing, 210 Little Falls Street, Church Falls, Virginia 22046. 12 issues p.a. Subscription US$65.00 (free to Society members). Articles vary in content from original scholarly contributions to short or concise technical notes. The views and opinions of members and authors are published, often provoking discussion and even controversy in successive editions. The remote sensing 'scene' that is conveyed is almost wholly that of the USA. In keeping with its origins, this journal continues to publish a significant number of papers specific to aerial survey, photogrammetry, and the interpretation of aerial photography. There is a special section listing academic and commercial job opportunities in most editions.

4. Remote Sensing Quarterly (formerly Remote Sensing of the Electromagnetic Spectrum) (Vol. 1, 1973). The journal of the Remote Sensing Committees of the Association of American Geographers and the National Council for Geographical Education, of specific interest to North American college and university geographers. Published by the Department of Geography and Geology, University of Nebraska at Omaha, Omaha, Nebraska 68182, £15.00 p.a.

5. SIEEE Transactions of Geoscience and Remote Sensing (Vol. 1, GE, 1962). American based with most editions containing articles of direct relevance to the application of remote sensing in the numerous sub-disciplines of earth science.

6. Washington Remote Sensing Letter (WRSL). An up-to-date source of news of international technical, educational, economic, and political developments that have a direct or indirect bearing on remote sensing activities. Contributions are mostly concise, and many are expressions of attitudes and opinions by individuals with decision-making or management responsibilities. Some articles are longer, and devoted to analysis of a specific product, process, facility, service, etc. Research, pure and applied, in academic institutions, public agencies, and commercial organisations is fully represented, with special emphasis on activity in the USA. All major reports, books, circulars, maps, and other materials are listed, together with a detailed calendar of forthcoming conferences, symposia, conventions, and other meetings. A valuable component is timely publicity for announcements of research contract and experimental opportunities offered by US federal government agencies. Published by Washington Remote Sensing Letter (WRSL), 1057-B National Press Building, Washington DC 20045, USA. 22 issues p.a. annual subscription is US$300.00 (1985).

7. Remote Sensing Society News and Letters which is sent free to all members appears four times a year, and contains a range of news information (some of it contributed by members and some from press releases and from entries in other journals) as well as short scientific papers. Also included are summaries of recent conferences, a calendar of forthcoming meetings in all parts of the world, and details of new educational courses. As would be expected, there is particular coverage of developments in the UK and West Europe, but developments in North America, Australia, and Third World countries receive attention. In most cases, contact addresses (sometimes names of individuals) are included, so that readers can follow up particular interests. The value of the papers (Letters) component is that the time between submission and publication is kept to a minimum. It cannot be obtained without membership of the Remote Sensing Society.

8. Other UK publications that contain a variety of news items are:

(i) Newsletter of the National Remote Sensing Centre which

is available free and on request to be placed on the mailing list.

(ii) Remote Sensing Observer, an informal newsletter produced primarily for staff of Imperial College, University of London, but available to other interested individuals. There is no subscription price. Details from the Editor, Remote Sensing Unit, Imperial College, Prince Consort Road, London SW7 2BB.

9. Photogrammetric Record (Vol. 1, 1969) two issues p.a. The Photogrammetric Society, London, Department of Photogrammetry and Surveying, University College London, Gower Street, London WC1E 6BT (free to members).

Many other journals carry articles and papers partly or wholly related to remote sensing, especially those devoted to earth and environmental science, geography, hydrology, agriculture, forestry, etc. The only way that these can be monitored and identified is through the use of abstracting and bibliographical publications. Journals published in the UK that routinely include papers of direct relevance to the subject include:

(i) Journal of the British Interplanetary Society (BIS) 12 issues p.a. Published since 1935.

(ii) Advances in Space Research Pergamon Press, Oxford and New York, four issues p.a. The official Journal of the International Committee on Space Research (COSPAR).

(iii) Weather Royal Meteorological Society.
Quarterly Journal RMS, Royal Meteorological Society.

(iv) Environmental Monitoring and Assessment D Reidel, Dordrecht; quarterly.

(v) Journal Land Resources Division. ODA, Tolworth.

(vi) Journal of Pattern Recognition; quarterly.

(vii) Satellite and Space Technology; monthly.

The British technical press also publishes worthwhile review, updating, and provocative articles alongside relevant news items, editorial comment, and book reviews. New Scientist (weekly) is especially recommended. Items of a more technological nature, and concerned with all types of satellite and spacecraft programmes and systems, are to be found in magazines devoted to aviation and aeronautical engineering, e.g. Aviation Week and Space Technology (USA) and Spaceflight (BIS).

Recently the Financial Times and Guardian have shown greatest awareness of remote sensing, and have included numerous news and feature articles.

Other overseas journals that provide a variety of well-selected and refereed articles, most of them specific to their country of origin, include:

(i) The Canadian Journal of Remote Sensing (Vol. 1, 1974). The major source of scholarly publications on all aspects of the development of remote sensing methods and application in Canada, though most issues contain articles of a general nature. Published annually by Canada Aeronautics and Space Institute, Saxe Building, 60-75 Sparks Street, Ottawa, Ontario KIP 5A5, Canada.

(ii) ITC Journal (No. 1, 1972). This is the main outlet for the publication of summaries of research carried out by the staff and students of the International Institute for Aerospace Survey and Earth Sciences (ITC) located at Enschede in The Netherlands. ITC is an essentially international teaching and training centre, strongly oriented towards the needs and interests of developing nations. For many years the institution concentrated on aerial survey, photogrammetry, and air photo interpretation, but since the mid-1970s it has expanded its terms of reference to incorporate the full breadth of remote sensing science. Journal articles reflect the origins and recent history of ITC and are both inter- and multi-disciplinary in approach. Topical coverage tends to be biased towards regional environmental resource surveys, methodological developments, and teaching/training approaches.

(iii) Photointérprétation (Vol. 1, 1962). Specific to interpretation from aerial photography, with a majority of articles in French. Published bi-monthly by Editions Technip, 27 rue Ginoux, Paris 75737. Annual subscription is 75f.

(iv) Photogrammetria (Vol. 1, 1946). Official journal of the International Society for Photogrammetry, with its principal editors working in Western Europe. Content is dominated by papers on the making of precise measurements on photographic images and the preparation and execution of map-making from aerial photos. However this journal publishes major papers on photo-interpretation relevant to a wide diversity of disciplines, from civil engineering to land cover assessment. More recently, coverage has embraced remote sensing in a broader sense. Authors and content reflect international perspective of the Society. Six issues p.a. Annual subscription is US$70.00. Published by Elsevier Scientific Publishing Co, PO Box 211, Amsterdam, The Netherlands.

(v) Remote Sensing Reviews (Vol. 1, 1981). Irregular, Harwood Academic Publishers, New York. International review articles of recent advances in both the theory and applications of remote sensing methodology. Edited in France.

(vi) The scarcity of reliable and substantial news, and of scientific papers in overseas journals, from remote sensing agencies, institutes, and individuals in the USSR is partly offset by the availability of the Soviet Journal of Remote Sensing (Vol. 1, 1981), published quarterly by Harwood Academic Publishers, Chur, Switzerland, at a subscription cost of US$380.00. The high cost of this journal reflects the economics of translation of papers appearing in Russian domestic publications. To an extent, the output of remote sensing specialists in East European States and Mongolia is also covered.

(vii) Aerial Archaeology (Vol. 1, 1977). Specific to the applications of vertical and oblique aerial photography and other imaging techniques in the identification and resolution of cultural remains. Content is biased towards North West Europe. Published by Aerial Archaeology Foundation, 17 Branston Square, London, and Committee for Archaeological Air Photography, 17 Bull Plain, Hertford.

Numerous other newsletter-style publications are issued on a regular or occasional basis by many national and a few international agencies, institutes, etc. Many are free or subsidised. Details are given under appropriate entries elsewhere in this guide.

A substantial technical and applications-related literature is contained in the published proceedings of conferences, workshop meetings, symposia, etc. held in many parts of the world. Inevitably, a good deal of repetition of basic ideas, themes, and techniques exists within and between these proceedings. They often contain draft editions of papers that appear later in expanded form, as refereed papers or reports and monographs. The collective theme of one conference may be only marginally different from another, so that titles alone lack distinction. Nonetheless, the time elapsing between the oral delivery of a paper and its subsequent publication may be as little as six months, thereby providing teachers and researchers with a good sense of the state of the art in given fields. They are a fruitful source for contact and exchange of expertise and experiences between individuals and groups working on closely-related problems in different countries. All of the major conference (etc.) publications are abstracted in detail by the bibliographical, indexing, and abstracting services previously noted. In some cases, certain publishers routinely publish the proceedings of conferences sponsored by single agencies, e.g. Plenum Press in the case of NATO Advanced Studies meetings.

It is impossible to note every set of proceedings here. Coverage is restricted to those that appear on a regular basis and have acknowledged international relevance.

1. International Symposia on Remote Sensing of the Environment. Published since 1962 by the organising agency, Environmental Research Institute of Michigan (ERIM), Ann Arbor, Michigan 48107. These annual conferences are ambitious attempts to draw together experts from a wide variety of disciplines, working in many parts of the world. They are normally sponsored by agencies, institutions, government departments, and international authorities with a commitment to remote sensing. There is usually restriction to a limited number of broadly defined themes, which change in successive years. Recently,

there has been the addition of periodic thematic conferences, e.g. Arid and Semi-Arid Lands (1981); Exploration Geology (1985). The main annual conference now meets in a different major world city every other year, e.g. Paris (1984) and Costa Rica (1980), alternating with Ann Arbor. The conferences themselves attract exhibitors from industry, commerce, and government, and are enhanced by displays, poster sessions, and excursions. The papers are presented in a series of parallel sessions, thus the proceedings are as vital to those who attend as to those who do not. However, the proceedings themselves cannot do justice to the varied range of activities and interests thus promoted. Individuals may request to be placed on the mailing list to receive advance information and calls for papers. The success of these conferences is guaranteed by the distinguished organising committee that is recruited for each event. As the proceedings for each symposium may run to over 2,000pp., a comprehensive index is essential. Two volumes of Indexes and Abstracts for 1962-1980 were published in 1981 (Vol. 1: Indexes; Vol 2: Abstracts), covering the 28 volumes of the first 14 symposia. It is an essential reference source for everyone engaged seriously in the application and teaching of remote sensing. The two volumes are priced at US$85.00, and may be ordered from Remote Sensing Center, ERIM, PO Box 8618, Ann Arbor, Michigan 48107. Supplementary indexes are in preparation for later meetings. It can be fairly stated that the 32,000pp. of the conference proceedings published to date provide a comprehensive documentation of the growth and maturation of remote sensing methods and techniques. Advances in all major application fields are fully represented. The quality of the index for 1962-1980 enhances this record still further by including keyword descriptors to categorise each paper and author affiliations.

2. Proceedings of the Annual Conference of the Remote Sensing Society: Each conference has a specific theme, which is reflected in the titles given below. Authors are responsible for copy, with only limited editing of the final record. One advantage of this approach is that volumes are available soon after the conference in question. The conferences themselves are the major event in the calendar of meetings for remote sensing specialists in the UK. In keeping with the Society's international outlook, speakers from overseas are strongly represented, and papers are drawn from global experience. Copies of past proceedings are available from the Society, c/o Department of Geography, University of Reading, Reading, Berkshire RG6 2AU at the prices quoted. Members receive a copy of the volume issued during the currency of their annual subscription.

Fundamentals of Remote Sensing (1974).
Microfiche only, £2.50.
Remote Sensing Data Processing (1975).
Microfiche only, £2.00.
Land Use Studies by Remote Sensing (1976)
Microfiche only £2.50.
Monitoring Environmental Change by
Remote Sensing (1977). £6.00.
Remote Sensing Applications in
Developing Countries (1978). £6.00.

Remote Sensing and National Mapping (1979). £6.00.
Coastal and Marine Applications of
Remote Sensing (1980). £8.00.
Geological and Terrain Analysis Studies
by Remote Sensing (1981). £9.00.
Matching Remote Sensing Technologies and
their Applications (1981). £12.50.
Remote Sensing and the Atmosphere (1982). £15.00.
Remote Sensing for Rangeland Monitoring
and Management (1983). £8.50.
Remote Sensing: Review and Preview
(10th Anniversary Conference) (1984). £16.00.

The volumes for recent years are substantially longer than the earlier ones. In addition, the Society has selectively published the papers presented at specially-convened short conferences and seminars. The full list of these proceedings is:

Proceedings of Meeting on Texture Analysis (1977). £2.00.
Proceedings of Seminar on Environmental Monitoring
by Remote Sensing (1979). Microfiche only. £2.50.
Remote Sensing and Strategic Planning (1981)
(with a bias towards Scottish experience), £2.00.
Field Radiometry (1983) £7.00.
Location..., Terrestrial Positioning and Geometric
Correction of Imagery (with University of Nottingham),
(1984), £2.50.

3. Proceedings of the William T. Pecora Memorial Remote Sensing Symposium. Annual, since 1975. Sponsored by the US Geological Survey, NASA and NOAA with the intention of obtaining an authoritative overview of recent and current advances in satellite sensor development and performances, data analysis, and applications to environmental problem resolution. Recent symposia have been extended to take in and evaluate changes in the administration of space remote sensing programmes and in market opportunities. Most papers are commissioned from leading exponents of new methodologies and those holding managerial posts in US government agencies. The proceedings as a whole may be regarded as a considered record of contemporary activity and thinking within the relevant US federal government offices, agencies, and contractors. Conferences are confined to space-based remote sensing programmes and products. Proceedings are published by the USGS Eros Data Center, Sioux Falls, South Dakota 57116 at an average price of US$15.00.

4. Proceedings of the Canadian Symposium on Remote Sensing (Vol. 1, 1972). Irregular, Canadian Aeronautics and Space Institute, Ottawa. Although the majority of papers derive from applications to Canadian environments and environmental problems, overseas work involving Canadian expertise is also represented. Technological innovations, and the well-structured, decentralised organisational framework that has been developed make the Canadian 'scene' of particular interest.

5. Proceedings of Annual Technical Symposium of Society of Photo-Optical Instrumentation Engineers. Each symposium

produces between 12 and 15 separate volumes, the whole constituting a complete record of the proceedings of the conference. Each volume may be bought separately at an average price of US$40.00. To date, nearly 350 volumes are available, listed in detail by the Society in a separate booklet. The majority of papers are concerned with the methods and techniques of optical, electro-optical, photographic, radar and laser technology, and their highly diverse applications in industry, medicine, and education. Only a relatively small proportion of contributions relate directly to the conventional field of remote sensing, with most of these specific to either the theory of electomagnetic energy transmission or to image and signal processing and analysis. The relevant papers are however often substantive and an important source of information on technological innovation. Details of the objectives of the Society, its continuing education programmes, publications, and conference plans may be obtained from SPIE, PO Box 10, 405 Fieldston Road, Bellingham, Washington, 98227. All symposia and other meetings are held in the USA.

6. Proceedings of the Annual Symposium on Machine-Processing of Remotely-Sensed Data (Vol. 1, 1974) Laboratory for Applications of Remote Sensing, Purdue University, 1220 Potter Drive, West Lafayette, Indiana 47906. A pioneering conference series devoted to the technology and methodology of digital image processing display, analysis and interpretation. Each conference attempts to assess the state of the art of the field (with most contributions coming from research groups in the USA) usually with reference to a defined application area. Symposia are always held at Purdue University, where delegates have the opportunity to attend customised post-meeting short training courses.

7. The Institute of Electrical and Electronic Engineers (IEEE) supports a Geoscience and Remote Sensing Society that holds a major conference every year that includes both invited and submitted papers. These are summarised in the Society's Digest, but many are published in the Transaction of IGARSS (Geosciences and Remote Sensing). The flavour of these meetings, and the scope of the papers, are international. Conferences are held in different major cities in successive years. Further details available from IEEE Service Center, 445 Hoes Lane, Piscataway, New Jersey 08854.

8. Archives of the ISPRS represents the record of the bi-annual congress, held in various centres, e.g. Hamburg (1980), Toulouse (1982), Rio de Janeiro (1984), organised by the International Society for Photogrammetry and Remote Sensing. The work of the Society is subdivided into several permanent Commissions, of which Commission VII (Interpretation of Photographic and Remote Sensing Data) represents the wider dimensions of remote sensing. The Commission delegates its terms of reference to discipline-specific Working Groups, e.g. Ice and Snow; Engineering Projects; Land-Use and Land Cover; Spectral Signatures; Methodology for Enhancement and Thematic Classification (of electromagnetic data). The focus and titles of Working Groups change according to consensus views on priorities and potentials. Conferences (and the published proceedings) of Working Groups are held at irregular intervals, not necessarily coinciding with the main congress. The

Archives are therefore an important record of international evaluations of the state of the art in specifically-defined areas. Up until 1980, the Society lacked the appendix of Remote Sensing in its title, and the Archives published prior to that date are more narrowly confined to photogrammetry, aerial survey, and photo-interpretation. The Secretariat of ISPRS is currently based at ITC, The Netherlands. Conferences are sponsored and supported by professional societies in various countries.

9. ITC Publications, Series A and B A long series of papers, most of them brief and devoted to a specific technique or case study. There are, however, a few conference and seminar proceedings, some produced jointly with FAO and UNESCO. The full list is available on request from ITC, at International Institute for Aerospace Survey and Earth Sciences, PO Box 6, 7500A Enschede, The Netherlands.

10. Remote Sensing Series (Vol. 1, 1980) Department of Geography, University of Zürich, Switzerland. Serial reports largely devoted to research carried out in Switzerland or on overseas contract. Agricultural and rural land use, soil moisture, and snow mapping are recurring topics of the titles in this series. Remote Sensing Section, Department of Geography, University of Zürich, PO Box CH-8033, Zürich.

E. REPORTS, MONOGRAPHS, INTERNAL AND OTHER 'NEAR PRINT' LITERATURE

There has been a tendency for the detailed results of remote sensing research to be written up in the form of agency and institution reports. They are often difficult to identify and may escape the attention of the abstracting journals. The majority of these publications come from the USA, including organisations such as National Oceanic and Atmospheric Administration (NOAA); National Aeronautics and Space Administration (NASA); US Army Corps of Engineers (Topographic Laboratories); Center of Earth Resources Management Applications (CERMA); US Geological Survey; US Department of Commerce; National Environmental Satellite Data and Information Service (NESDIS); Technology Applications Center, University of New Mexico (TAC); Environmental Research Institute of Michigan (ERIM); Jet Propulsion Laboratory, California Institute of Technology (JPL); Laboratory for Applications of Remote Sensing, Purdue University (LARS); National Space Technology Laboratories (NSTL); University of Kansas Center for Research; and National Technical Information Service (NTIS). Most of these organisations publish lists and indexes of their reports, and it is fortunate that NTIS collates and indexes many of these publications as well as publishing bibliographic reviews and other memoranda in its own right. All NTIS indexed reports are available directly through the UK repository, the Technology Reports Centre. Microfiche copies are normally supplied.

An important, and fully documented, information source on all NASA reports is contained in Scientific and Technical Aerospace Reports (STAR) produced by the agency's Scientific and

Technical Information Branch. It is published twice monthly and constitutes a complete record of all reports produced by NASA staff and by contractor and grantee organisations. It also includes reports of relevance from other US and foreign institutions, and commercial companies. Each entry is accompanied by an abstract and accession details. Monthly index issues are also published. Books, conference proceedings, journal articles, and translation of foreign language reports are abstracted in a companion publication, International Aerospace Abstracts (IAA). STAR covers recently completed theses from universities and other institutions and carries a section on NASA-sponsored research in progress. It is available on subscription from the USGPO, Washington DC 20402 (US$150 p.a., for overseas addresses, in 1983). Microfiche of abstract sections of each issue (retroactive to 1974) are supplied by the NTIS (5285 Port Royal Road, Springfield, Virginia 22161) at an annual subscription slightly higher than for hard copy. A booklet describing the NASA Scientific and Technical Information Service is available on request from NASA, STIS, PO Box 8757, BWl Airport, Maryland, 21240.

A large number of NASA reports (though not all) have been deposited in the British Library, Lending Division, Boston Spa, Wetherby, Yorkshire, and are available through the national inter-library loans service. The British Library also holds copies of reports from many other North American sources. Microfiche copies of all NASA and NASA-sponsored reports are automatically received by the Information Retrieval Service, European Space Agency, ESRIN, Via Galileo Galilei, 00044, Frascati, Rome, Italy.

As the agency responsible for the operational control of the majority of US satellites, rather than research and development, the output of report literature from NOAA is relatively small. There is, however, a Technical Memoranda series, available from NTIS or Enviromental Data and Information Service (EDIS), 6009 Executive Boulevard, Rockville, Maryland 20852. EDIS is a constituent agency of NOAA.

Other national remote sensing units and centres collectively publish a substantial report literature. These are less easily traced, although the 'Reports' section of the former RSS Newsletter (up to 1983) noted many of them. The same applies to reports from inter-governmental organisations such as Food and Agriculture Organisation (FAO), Rome; World Meteorological Organisation (WMO), Geneva; UNESCO (Paris); European Space Agency (ESA), and others. Direct correspondence with any of these institutions will normally produce details of past and current publications, and most of them have a library or documentation service. Technical Reports and other papers published by the Canada Centre for Remote Sensing are a particularly useful insight into the relevance of satellite remote sensing in a large, sparsely-populated national territory where difficult or hazardous terrain conditions inhibit direct survey, mapping, and monitoring approaches. CCRS has, in addition, made a significant contribution to the development of microwave, notably SAR, remote sensing and to digital analysis. Up-to-date news of contemporary developments in Canada is contained in a regular newsletter

Remote Sensing in Canada published free of charge by the CCRS, 2464 Sheffield Road, Ottawa K1A OY7, Canada. This publication also summarises or digests recent news on remote sensing in North America as a whole.

Reports from various sources (mostly specific to Landsat) are selectively abstracted in Landsat Data Users Notes from NOAA, Landsat Customer Services, Mundt Federal Building, Sioux Falls, South Dakota 57198, free of charge, four times a year.

Technical reports from sources in the UK are not, as yet, sufficiently numerous to escape notice in newsletters and journals published in Britain or by abstract services. Readers of any of these publications have the reasonable guarantee that new reports will be drawn to their attention. Sources of reports, circulars, brochures, etc. include the NRSC and its Working Groups, Department of the Environment, Natural Environment Research Council (NERC), Science and Engineering Research Council (SERC), Institute of Oceanography (Godalming), Institute of Hydrology, Soil Survey of England and Wales, Meteorological Office, The Robert Hooke Institute (Oxford), British Geological Survey (including Overseas Geology Unit), Department of Trade and Industry, Institute of Terrestrial Ecology, Nature Conservancy Council, Ministry of Agriculture, Fisheries and Food, Royal Commission of Historical Monuments, Forestry Commission, the Overseas Development Administration, Transport and Road Research Laboratory, Warren Springs Laboratory, and commercial companies (listed elsewhere).

Commercial companies also produce internal reports and publicity material. Universities, polytechnics, institutes of higher education and other tertiary educational establishments publish reports, usually as part of an occasional papers series or similar format. Those institutions with teaching and/or research commitments to remote sensing, listed under Educational Establishments, are the most likely sources of papers of this nature.

An early set of reports that might otherwise be missed is the Remote Sensing Papers and Remote Sensing Reports (12 titles in total) from the Department of Geography, University of Reading, published between 1973 and 1975. All are specific to the evaluation of the results of a Skylab Rocket Photography project carried out over the Pampas of Argentina. The use of this type of platform is unusual, and the experience reported records a specifically British contribution to remote sensing from space.

Higher degree theses constitute the most detailed record of research carried out in educational establishments, though they are often summarised in one or more papers published subsequently in scientific journals or conference proceedings. The easiest way to trace titles is to use the documentation service provided by University Microfilms International. 30-32 Mortimer Street, London W1N 7RA. A list of reports and monographs on microfilm/microfiche is also available on request.

Selective lists of theses completed in US universities are published from time to time in Photogrammetric Engineering and

Remote Sensing starting with Vol. 45 (Part 5) (1981), pp. 617-629. There is no updated directory of theses on remote sensing submitted at British universities and polytechnics. A NRSC-sponsored survey and report on post-graduate research activity, with a retrospective list of theses, is in active preparation at the time of writing.

The Remote Sensing Yearbook, Taylor and Francis, London. The first edition, published in 1985, includes details of current research and recently published theses as well as other technical literature orginating in the UK.

PART 2.3: AUDIO-VISUAL RESOURCES

A. SLIDES, SLIDE/TAPES AND OVERHEAD PROJECTOR TRANSPARENCIES

Emphasis is given to products from British publishers, which were in print at the time of writing. The order in which they are given does not imply their relative merits.

UK BASED SUPPLIERS

1. National Remote Sensing Centre

A set of 18 35-mm slides covering the physical basis, image types, aspects of image processing and interpretation, and some examples of applications. Some of the slides combine all of these attributes, and each is discussed in a detailed and authoritative set of notes ideally suited to introductory courses in higher education. Created by the Education and Training Working Group of the National Remote Sensing Centre and published in 1982, the set costs £6.85; all these can be obtained from the Remote Sensing Sales Centre, 172 King's Cross Road, London WC1X 9DS (01-278 8276). A set of 20 slides of Landsat TM imagery, produced by EARTHNET (1985) can be ordered via the NRSC.

2. Woodmansterne Limited

Six sets, each of nine slides consisting mostly of true colour obliques from manned spacecraft in Earth orbit obtained in the late 1960s and early 1970s are available under the umbrella title Earth from Space. There are several individual titles: Africa, Asia, Europe, America North and South, and America: USA. Full details can be obtained from Woodmansterne Limited, Watford, WD1 8RD (0923 28236).

3. Focal Point Audio visual Limited

(i) The Interpretation of Remotely-sensed Images This is a set of 40 slides and an accompanying tape of 45 minutes duration covering the basic physical principles of remote sensing and includes a variety of image examples. A slight majority of the latter include satellite-based imagery, with the remainder devoted to lower altitude platforms operating in different parts of the electro-magnetic spectrum. Emphasis is on examples from the British Isles, with senior secondary and junior undergraduate students of geography as the target audience. Published in 1981, the set costs £20.13.

(ii) The Earth From Space series consists of 12 sets of Landsat false-colour composite images of large regional areas. Each set contains 20 slides, with the exception of that of the British Isles which extends to 40 selected examples. Each production is accompanied by a booklet

LIVERPOOL INSTITUTE OF HIGHER EDUCATION THE BECK LIBRARY

describing and interpreting each image in a style most appropriate to teachers of geography and environmental studies. The titles are: Europe; British Isles, Northern and Western Africa; Southern and Eastern Africa; Southern and South-East Asia, China and Japan; Middle East; Canada and Alaska; USA - Central; USA - Western; USA - Eastern; Latin America and the Caribbean region; Australasia; the USSR. Most of the examples are of Landsat 1 and 2 images obtained between 1972 and 1978. Published in 1980-1981, each set costs £9.50, £15.80 for the British Isles set. Preview details are available which list briefly the content of each slide.

(iii) Voyager in Space: Jupiter and Its Moons; Voyager in Space: Saturn Published in 1982, these consist of 30 colour slides of some of the most informative and visually exciting images of Voyager's encounters with the two giant planets of the solar system. There is an emphasis on planetary atmospheres and gross geomorphology. The packages cost £9.50. Further details of all these products can be obtained from: Focal Point Audio Visual Limited, 251 Copnor Road, Portsmouth PO3 5EE (0705 665249).

4. Space Frontiers Limited

A variety of sets, most of them consisting of twelve or twenty slides, covering a range of topics within the general field of space science and exploration. Each set is accompanied by full, expertly-written explanatory notes. Titles of relevance to terrestial remote sensing are as follows:

Places: The World from Orbit (1978); Geology from Space: Plate Tectonics (1980), written in conjunction with the Association of Teachers of Geology; The United Kingdom from Space (two sets, 1975 and 1976); The Third Planet: Deep Space Views of Earth (1978); Meteorology from Space (1981), published jointly with the Royal Meteorological Society.

Slide sets of wider relevance include Images of Space; Man in Space; Skylab (two sets); Viking on Mars; Jupiter and Saturn: The Missions of Voyager 2 and Pioneer; Mercury, Venus and Jupiter: Images from Mariner. Space Frontiers market charts and posters (which are listed later in this section) and maintain an extensive library of original space imagery for use by publishers and other media. This is administered by The Daily Telegraph Colour Library, 135 Fleet Street, London EC4A 4BH (01-353 4242, Ext. 3686/3687/3688), to whom initial enquiries should be addressed. Further details can be obtained from: Space Frontiers Limited, 30 Fifth Avenue, Denvilles, Havant, Hants, PO9 2PL (0705 475313). Copies of slide sets are available only whilst stocks last as reprinting is not planned unless there is a specific demand.

5. Aerofilms Limited

An extensive library of oblique and vertical air photos of the British Isles is available as 35-mm transparencies. The high quality oblique views (in both black and white and colour) that compose the two series Flight Round Britain and Educational Sets are available either collectively or individually as slides. Without exception these slides provide the teacher of geography or geology with stimulating illustrative material. OH transparencies of many of the vertical air photos available in Aerofilms Educational Sets can be purchased in association with the print copies or individually. Further details of the entire collection of subjects, together with prices, can be obtained from: Aerofilms Limited, Gate Studios, Station Road, Boreham Wood, Herts, WD6 1EJ (01-207 0666).

6. Other Commercial Survey Companies

(i) Clyde Surveys

Single or multiple copies of 35-mm transparencies of all photographic prints available in the archive of the Landsat Products Service can be provided on request. The archive reflects the company's involvement in a wide variety of projects overseas, particularly in Africa and the Middle East. General enquiries to: Clyde Surveys, Reform Road, Maidenhead, Berks, SL6 8BU (0628 21371).

(ii) Nigel Press Associates

Similar to (i), based on an extensive image library. Full details can be obtained from: NPA Limited, Old Station Yard, Marlpit Hill, Edenbridge, Kent, TN8 5AU (0732 865023). A list of available slides will be supplied on request.

7. Meteorological Imagery Slide Sets

(i) Meteorology from Space was published in 1981 by Space Frontiers Limited (see 4 above); it consists of 12 slides. It is available directly from the Royal Meteorological Society, James Glaisher House, Grenville Place, Bracknell, Berks, RG12 1BX (0344 422957).

(ii) Weather Study with Satellites is a series of three slide sets published in 1982, each consisting of 38 frames and accompanied by detailed notes written by R. Fotheringham and R. S. Scorer. These provide comprehensive treatment of observation of atmospheric phenomena by orbital weather satellites. The sets are also available as filmstrips, and each set has an optional accompanying cassette providing a more concise commentary by Jack Scott, former Senior Meteorological Office weather forecast presenter for the BBC. The titles of the individual sets are as follows: Part 1 - Cyclones, Fronts and Anticyclones; Part 2 - Waves, Stratus, Fog and Sea Breezes; Part 3 - Convection Patterns, Vortices and Global Views. The slides with notes are £17.25 per set, the cassettes £4.60 each, the cassette texts 50p each and the filmstrips £11.50 for each part. Further details are available from: Diana Wyllie Ltd, Unit 8, Deverel Drive, Grandy, Milton Keynes MK1 1NL (0908 642323).

(iii) Observing and Forecasting the Weather comprises 20 slides and full, accompanying explanatory notes with suggested questions for further study with student groups. Somewhat over half of the slides are of imagery or data displays obtained by remote sensing methods. The remainder are concerned with observation techniques and applications of data (e.g. ship routeing; rainfall probabilities; agricultural forecasts, etc). The notes were written by Dr R Reynolds, and published in 1983 by Focal Point Audio Visual (see 3. above).

8. Livingston Studios

Livingston Studios distribute a filmstrip and cassette of a 20 minute broadcast from the BBC Schools Advanced Studies: Geography Series, originally transmitted in May 1982; it is a general overview of the capabilities and potential of global remote sensing, and is titled Environmental Monitoring: The Challenge of Remote Sensing. This original Radiovision production is in the form of a discussion with Dr A J Allan and D J Carter and is targeted to A-level geographers. The filmstrip contains 12 frames, with a deliberate bias towards Africa. It is available from Livingston Studios, Brook Road, Wood Green, London N22 6TR (01-889 6558).

9. Armagh Planetarium

A wide variety of both short and longer sets of imagery of planetary bodies in the solar system visited by American missions in the 1970s and 1980s is available. Each set is accompanied by brief notes. The sets (of 20 colour slides) titled Saturn Encountered and The Solar System are remarkably good value at £4.00. Details of titles (which include slides marketed on behalf of other organisations) can be obtained from: The Planetarium, College Hill, Armagh, Northern Ireland, BT61 9DF (0861 523689).

10. John Wiley & Son Limited

A fully documented collection of 276 early Landsat 1 images of geological and geomorphological features and processes entitled Earth Perspectives and reproduced as 35-mm slides is available. The majority are colour composites. Published in 1975, the collection is available from: John Wiley & Son Limited, Baffins Lane, Chichester, PO19 8JL.

OVERSEAS SUPPLIERS

1. NASA and EROS Data Center

Sets of slides, as well as individual copies of specific images, are available directly from NASA and the EROS Data Center. Details of this service will be provided on request.

2. Pilot Rock Inc.

An extensive collection of sets of 35-mm colour slides specific to a wide variety of teaching topics is available from Pilot Rock Inc. The majority of examples are taken from the landscapes of North America, with full use of low, medium and high altitude vertical and oblique air photos as well as satellite and spacecraft-derived imagery. Teachers in the UK would find any one of these sets a suitable source of examples to represent the geographical processes and patterns characteristic of North America.

The list of titles is too long to reproduce here, and a copy of the full catalogue should be requested. A selection of available sets includes the following titles:

Infra-Red High Altitude Photography; Urban America; Lakes and Reservoirs; Arid Landforms; Fault Features; Forest Resource Analysis; Techniques of Land Use and Management; Hydrology and Dynamics of Landforms in the Coastal Zone; The Synoptic View of World Food Resources and Agriculture; An Application of Remote Sensing in Geology and Geological Hazards; Volcanism; Introduction to Digital Processing of Landsat Data, and Applications of Remote Sensing in Archaeology. Explanatory and background notes accompany each set, many of which have been compiled and written by well-known researchers and authors. Some sets are specific to advanced level students, such as the Signature series devoted to applications of remotely-sensed data in environmental resource analysis. Others are of a more introductory nature and are adaptable to a variety of teaching levels. The company's Visual Teaching Aids brochure indicates clearly the target audience of each set. There are over 15 sets in the Regional series, incorporating a selection of imagery of major regional areas of the conterminous US, Alaska and Hawaii. Special teaching kits containing maps, notes and exercises as well as original imagery have been published since 1981, and are noted in more detail in Part 2.5.

Most of the sets consist of 20 slides with a few incorporating 25 or 30. A large file of over 35,000 images is available for individual reproduction as slides or prints according to systematic or regional interests. The address of Pilot Rock Inc. is: PO Box AS, Trinidad, California, 95570, USA.

3. Technology Application Center (TAC)

TAC distribute annotated sets of slides, varying between ten and 50 in a single collection which are particularly strong on Skylab and earlier Gemini and Apollo true colour photographs. Seventeen sets are devoted exclusively to Skylab products, arranged systematically under headings such as Geomorphic Provinces, World Deserts, US Agriculture, World Cities, Environmental Quality, Land Use and Oceanography. There are also some unusual regional sets e.g. Mainland China and Australia/New Zealand. One set of 40 slides consists of Skylab images of Europe. Since 1975, The TAC Audiovisual Library of Remote Sensing and Space Technology has generated a series of tape-slide programmes designed to introduce specific topics (especially thematic applications) to students and more general audiences (e.g. briefing meetings for policy-makers, engineers, etc). The emphasis of the series is on application fields. Each programme consists of a commentary of about 17-minutes duration accompanied by 50-60 slides; a filmstrip version is available as an alternative to the framed slides. At a rate of almost four slides or frames per minute, teachers will seek the opportunity to use them in other contexts. They will probably choose to use the commentary in short sections to give more opportunity to relate to the image details. Some of the programmes are accompanied by a study or discussion guide, or by a fact sheet, but the purpose of these slide/tapes is that they should be 'free-standing'. Some of the titles include: Food Watch by Satellite; Hydrology by Satellite; Remote Sensing and Economic Development; Wildlife Management and Remote Sensing; Forestry and Remote Sensing, and Aerial and Orbital Systems for Remote Sensing. Further details and prices can be obtained from: The TAC Library, Audio-Visual Institute, 6839 Guadalupe Trail NW, Albuquerque, New Mexico, 87107, USA (0101 505 256 0808).

4. Remote Sensing Enterprises

An extensive collection of 35-mm slides on a range of topics is available from Remote Sensing Enterprises, PO Box 2893, La Habra, California, 90681, USA. They have been selected and the accompanying notes written by Professor Floyd Sabins. The target audience is that of undergraduate and postgraduate students. Taken as a whole the collection provides a comprehensive audio-visual resource for any course in remote sensing. Of the several titles, some may be mentioned, viz Comparison of Image Types (30); Thermal Infrared Imagery (49); Aerial Photographs (35); Radar Imagery (50); Resource Exploration (59) and Natural Hazards (30 slides). One set (100 slides) titled Remote Sensing Laboratory Manual and Instructor's Key consists of reproductions of images, diagrams, tables, maps, etc. appearing in the same company's Remote Sensing Laboratory Manual and Instructor's Key for Remote Sensing Laboratory Manual. The three products are therefore complementary and provide extensive scope for theoretical, practical and discussion work. All of the exercises are the product of careful evaluation based on the author's extensive field, consultancy and teaching experience; the slides give the opportunity for either detailed commentary and feedback from teachers or for forms of self-assessment by students. An innovative and attractive arrangement is the publication of a separate Guide and Explanation for Remote Sensing Slide Sets, in which each slide (in every set) is described in 100-200 words. This Guide is available separately in advance of any purchase. The wealth and variety of image types and subjects, and the incorporation of worked interpretation exercises with obvious application contexts, make for a flexible, high quality teaching and learning resource. Further details and current prices can be obtained from the above address.

5. Purdue University, Laboratory for Application of Remote Sensing (LARS)

Perhaps the most well known, and widely adopted, audio-visual resource in remote sensing teaching at the tertiary level is the Minicourse series of slide/cassette programmes, under the umbrella title of Fundamentals of Remote Sensing. It was designed and developed by a team of LARS staff that included theoretical scientists, engineers, specialists in application fields and teaching staff. Each Minicourse is a self-contained teaching package that can be used for independent student learning programmes or for integration into formal class presentations; each one can also be used as library resource for student use in consolidating understanding.

Each package consists of an average sequence of 30 slides and an accompanying pulsed tape commentary of about 25 minutes duration. At regular intervals questions are posed and exercises suggested. The detailed instructions for these are given in an accompanying booklet or study guide which also reproduces (in black and white) imagery on the slides that are used in interpretational and analytical questions. The booklets also serve as a summary transcript of each spoken commentary. Exercises serve to consolidate understanding or exemplify particular points; they can be undertaken by student groups as they listen to the tape, or they can be attempted afterwards. The answers to single and direct questions are given on the tape; 'model' interpretations and answers to computational questions are given in the accompanying Instructor's Guide. In each pack, there are 25 copies of the study guide, sufficient for the average-sized group that would use this resource. Most of the study guides include additional self-tests and reproduce statistical tables or more complex maps and diagrams that appear on the slides.

This is obviously a product for relatively advanced students, although the levels of assumed background awareness vary considerably. To some extent, the Minicourse units devoted to broadly defined topics are prerequisites for those of a more specific and specialised nature. Teachers will find that the Minicourse format is not a limitation to their use; many of the slides and activities (exercises) can be extracted and used in different learning contexts.

The first titles were published in 1976, with several others added in 1981. A full list is available on request together with a descriptive booklet and prices. A comprehensive overview of the entire series, the Instructor's Guide, can be purchased, which gives a summary, objectives, prerequisites and details of equipment needed for each Minicourse. The guides accompanying individual packages include contact prints of each slide and master copies of specific components such as maps, OHT overlays, etc.

Some of the titles include: Physical Basis of Remote Sensing; Sensor Systems; Numerical Analysis; Pattern Recognition in Remote Sensing; Typical Steps in Numerical Analysis, Multispectral Scanners; Spectral Reflectance Characteristics of Earth Surface Features; Principles of Photointerpretation; Interpretation of Radar Imagery; Interpretation of Thermal

Imagery; Skylab: Earth Resources Experiment; Application of Remote Sensing in Geology; Crop Surveys Through Remote Sensing; Temperature Mapping of Water; Mission Planning: Considerations and Requirements; Mineral Exploration Using Satellites. The Minicourse on Selecting Landsat Imagery was developed in cooperation with ITC, Netherlands. Full details, prices and orders to: Continuing Education Administration, 116 Stewart Center, Purdue University, West Lafayette, Indiana 47907, USA (0101 812 494 4600).

6. The International Training Centre for Aerial Survey and Remote Sensing

The ITC publishes three series of audio-visual products, all of which are relevant to advanced teaching and instruction. The first is the Hothmer Stereo Slide Collection, a total of 86 5 x 5 cms (2" x 2") slides for stereo-projection and viewing. In every case two identical images are mounted on to one slide and carefully calibrated to ensure stereo vision, which requires a special attachment to standard projectors, or polarised spectacles. ITC issue a list of suitable projectors and attachments, all of which are marketed worldwide. The slides are arranged in a number of series, viz. Geology (36); Geography (9); Forestry (13) and Soil Science (28). The emphasis of each collection, detailed in the accompanying notes, is on feature and pattern identification and interpretation. The scenes are all of air photographs at original scales of between 1:10,000 and 1:50,000. The second series is a long one of over 600 slides, arranged systematically, illustrating photogrammetric instruments, aerial survey cameras and non-photographic sensors. Some of these slides are also available for stereo vision. The third series is a set of slide/cassette tape presentations published in 1981. They have an average playing duration of 35 minutes and are based on a sequence of about 35 slides. The titles currently available are: Slotted Templates; OM1 Stereo Facet Plotter; Stereo Zoom Transfer Scope; Airphoto Interpretation for Rural Land Use (65 slides and accompanying exercise workbook); Airphoto Interpretation for the Study of Rural Areas (one hour tape track and 154 slides); Fundamentals of Remote Sensing (also in the Purdue LARS collection). An additional one hour tape with 149 slides and a sequence of exercises is devoted to The Basic Principles of Multicoloured Map-Making, which is only marginally relevant to remote sensing in the strict sense.

ITC is also developing a Videotape series, of which the first two titles (Air Survey Mission and How to use the Diatype) are now available. Both are of 25 minutes duration.

Further details of authorship, content approach and prices can be obtained from: ITC, Boulevard 1945, AA 7705 Enschede, Netherlands.

7. Spot-Image

In anticipation of the launch and operation of the SPOT satellite, the company responsible for all aspects of access to

SPOT data and marketing of services and products has published a 20 slide set, Examples of Spot Simulation, using images derived from a variety of airborne simulation campaigns carried out by investigators in different parts of the world, as well as in France. There is also a set of 18 Technical Sheets outlining SPOT technology and thematic applications (e.g. cartography; crop statistics; land use; town planning in Paris; coastal environments) that may be purchased as two separate folios. Prices and further details can be obtained from: SPOT-IMAGE, 18 avenue Edouard-Belin, 31055, Toulouse, Cedex, France, or their agents in the UK: The National Remote Sensing Centre in Farnborough or Nigel Press Ltd. All products are available in both French and English language editions.

8. Gregory Geoscience Limited

A 22 slide set of Landsat colour composite scenes of Canada was published in 1976 by Gregory Geoscience Limited (1750 Courtwood Crescent, Ottawa, Ontario, K2C 2B5, Canada). The accompanying descriptive notes are exceptionally detailed for this type of product and refer to numbered annotations on the projected slides. The collection as a whole is representative of the variety of terrain and land-use conditions in Canada (so far as such a vast national territory allows this). This is a flexible product, especially useful in geography and geology teaching at elementary through to senior student levels. There is, however, little acknowledgement of image enhancement, as most of the slides are standard composites using various Landsat band combinations. The price of the set is $C22.50.

9. Nelson Audio-Visual

A set of 18 overhead transparencies of early Landsat false-colour composites, originated by Westermann (West Germany) is marketed by Nelson Audio-Visual. Each transparency is accompanied by an overlay with annotations of salient geographical features, place-names and image orientation, and teacher's notes providing background regional description. The notes also suggest exercises that could be used with CSE, O level and first year A level geography students. They emphasise basic skills of identification and interpretation, and encourage the use of ancillary sources such as atlas maps. Published in 1979, the notes are written to relate to the 'core' content of school geography syllabuses and are therefore adaptable. The image examples all show strong, definite patterns and regional boundaries defined by both physical and cultural features, such as the Dakhla Oasis of Egypt; southern and central California; Harz Mountains, West Germany; Banff region, Alberta (Canada) and the Dutch polders marginal to the Ijsselmeer. Each image and accompanying materials are separately packaged and available for sale individually at £4.04. In addition, Nelson have also adopted a Westermann 24 frame full colour filmstrip of Landsat false-colour scenes of examples of major cities and coastline types. The twelve coastal examples are taken mostly from Europe, but the coverage of urban centres is worldwide. Explanatory notes in the form of a booklet, acompanying the filmstrip, which costs £8.63, can be obtained from: Nelson Audio-Visual, 51 York Place,

Edinburgh, EH1 3JD (031-557 3012).

B. FILMS AND VIDEO PROGRAMMES

With the exception of the extensive catalogue of films availabe from NASA, there is at present a strictly limited number and variety of relevant titles. Most are of a general, overview nature suitable for introducing the subject to pupils and students. Most of them are somewhat out of date in terms of technological development. In the listing below, (S) indicates suitability for secondary school pupils; (H) suitability for students in higher and further education.

Several of these films are available on free hire, but costs of carriage are normally paid by hirers. Conditions of hire may be obtained in advance from distributors.

1. Films

NASA Films(S) (H). An up to date catalogue of all currently available films is available from NASA. They are available on free loan from the various regional centres in the USA. In the UK, the exclusive distributors are: The British Interplanetary Society, 27/29 South Lambeth Road, London SW8 2SZ (01-735 3160). A duplicated list of available titles, together with abstracts of content and details of hire conditions and charges, is available on request. This list is a representative rather than complete collection of NASA output. Many of the titles relate to the engineering and biomedical aspects of space research as well as to extra-terrestrial manned and unmanned missions. A series of short 15-minute films was made in 1975-6 relating to the potential of Landsat for environmental mapping and monitoring, but these are now in need of revision in view of the subsequent ten years of experience of operating this programme. Of somewhat more value is a film, released in 1977, titled Growing Concerns which indicates how the Landsat programme can provide timely surveillance of forest and field crops; the first results of project LACIE are included.

The titles in the series noted above are: Earthquake Below (specific reference to San Francisco); Tornado Below; Flood Below (very little content directly related to remote sensing); Pollution Below; The Pollution Solution; The Fractured Look; The Wet Look and Remote Possibilities (the last title defines remote sensing and takes a wide spectrum of examples of Landsat's possible contributions to the management of natural resources).

NASA films can be directly purchased from the National Audio Visual Center, General Services Administration, Order Section, Washington DC 20409, USA. Orders from outside the USA must be accompanied by payment.
Higher Education Film and Video Library acts as distributor for a number of film-producing companies and public agencies. Booking requests should be addressed to: HEFVL, SCFL, Dowanhill, 74 Victoria Crescent Road, Glasgow, G12 9JN (041-334

9314). Relevant titles include:

Minerals Exploration: The Use of Remotely-sensed Data, 1980, 27.5 mins (H).

Vegetation Assessment: The Use of Remotely-sensed Data, 1980, 27.5 mins (H).

These are companion films, originating from the American Society of Photogrammetry.

Remote Sensing: A New Angle on the World, 1980, 30 mins. An English language edition of an orginal film produced by the French Scientific Film Service in close co-operation with the Institut Géographique National, Paris. It achieves a balanced and well-integrated overview of the principles, technology and application potential of remote sensing. The context is exclusively that of French research and development (S), (H).

Remote Sensing by Satellite: New Dimensions to Our Knowledge of the Environment, 1981, 17 mins, French Scientific Film Service. Somewhat similar to above title, but condensed and orientated more strongly towards the user community through the use of a regional case example of the use of Landsat multispectral data in West Africa. The commentary is entirely non-technical (S).

Food Production Systems: Monitoring, 1979, 24 mins. An Open University/BBC Production made as an integral part of an OU-distance learning course on food production. Available as either 16-mm film or VHS video from Guild Sound and Vision, Woodston House, Oundle Road, Peterborough PE2 9PZ (0733 63122). The subject concerns an evaluation of satellite remote sensing as a means of carrying out crop censuses in developing countries where conventional methods of data acquisition may not be available (H).

The Long View, 1973, 22 mins. National Film Board of Canada. Available on hire from Canada House Film Laboratory, 1 Grosvenor Square, London W1V 0AB (01-629 9492). Made in the early days of the Landsat programme and therefore very dated from that point of view, this film nonetheless achieves a balanced overview of the physical basis of remote sensing (S).

Promise in the Sky, 1980, 29 mins. Produced by the Central Office of Information, on behalf of the Foreign and Commonwealth Office; available from: Central Film Library, Chalfont Grove, Gerrards Cross, Bucks, SL9 8TN (02407 4111), Cat. No. UK3442. The context of this film is somewhat wider than remote sensing, as it includes telecommunications satellites and their role in facilitating international exchange. Nonetheless the relevant content is provocative and supported by an instructive case example. Surprisingly, British involvement in remote sensing is not specifically featured (S) (a new edition of this film is due for release in late 1985).

Photography and the City, 1969, 15 mins. Distributed by Concord Films Council Limited, 201 Felixstowe Road, Ipswich, Suffolk, IP3 9BJ (0473 76012). The film comprises multiband camera, thermal infrared and radar imagery of selected American cities, obtained from both aircraft and satellites employed to show historical and contemporary land-use change. Computer graphics based on processed digital data highlight certain spatial characteristics (H).

Films and videos based on both polar orbiting and geostationary meteorological satellite imagery are available from: CDZ Film Berlin, Schluterstrasse 39, 1000 Berlin 12, Federal Republic of Germany (30 882 7557), from whom a catalogue of the wide selection of titles may be requested. Some of the productions are on behalf of the European Space Agency, and all are available with English commentaries. Themes range from global scale circulation patterns to tertiary scale features such as sea-breezes. Some of the films contain animated time-lapse sequences (H).

A further selection of films and videos (S), (H), all of them based on American geostationary meteorological satellite imagery, is produced by: Walter A Bohan Co, 2026 Oakton Street, Park Ridge, Illinois 60068, USA.

2. Videos

Remote Sensing: Mineral Exploration (H) (43 mins, 1980), Remote Sensing: Coastal and Marine Applications (H) (15 mins, 1985) and Synthetic Aperture Radar (60 mins, 1981) (H) are commissioned or 'in house' productions of the National Remote Sensing Centre suitable for higher education students. (The target audience for the two tapes on radar is that of post-graduates and others with specialist interests in the field of microwave remote sensing). Both titles have been produced in close co-operation with practising experts and concentrate on the techniques, and value, of digital image processing and enhancement. Examples are taken from research projects in various parts of the world in which the NRSC and contributing agencies have had involvement. The films are available for purchase as VHS or UM format from NRSC, Space Department, Royal Aircraft Establishment, Farnborough, Hants, GU14 6TD and soon from the Remote Sensing Sales Centre, 122 King's Cross Road, London WC1X 9DS. Geology from Space, an Open University/BBC Production (17 mins, 1984) (H) available from ERSAC Limited.

New Techniques in Image Processing (28 mins, 1981), distributed by the University of London Audio-Visual Centre, 11 Bedford Square, London WC1B 3RA (01-387 7050) and produced by the Laboratory for Planetary Atmospheres (Imperial College, formerly at University College) illustrates the capabilities of their Interactive Planetary Image Processing System (IPIPS). Exemplification is provided by Voyager imagery of Jupiter and Saturn, including some time-lapse sequences and Meteosat scene rectification (H).

Landsat and Satellite Weather Watchers (17 and 12 mins, respectively, 1982) produced by the Australian Department of Science and the Environment and available as UM (only) from Darvill Associates, 280 Chartridge Lane, Chesham, Bucks, HP5 2SG (0494 783643): the first reviews how Australia has, and should, benefit from space remote sensing, whilst the second film features the Japanese Geostationary Meteorological Satellite (GMS), positioned over West Irian (S) (H).

Food Production Systems - see description of this on p. 105, is also available as a video cassette (H).

A selection of NASA films is available as video cassettes from Istead Audio-Visual, 38 The Tythings, Worcester (0905 29713). A descriptive brochure is available from the company, who act as sole UK agents for this format (S) (H).

The US National Space Science Data Center acquires videos, for hire, from various American public agencies. Overseas requests should be addressed to the World Data Center A for Rockets and Satellites, Code 601, Goddard Space Flight Center, Greenbelt, Maryland 20771, USA. Recent additions include JPL productions on the Shuttle Imaging Radar, which show terrain features as they would be viewed by an observer flying on board the Shuttle.

Purdue University: in addition to the series of slide/cassette programmes in the Minicourse series produced by the Laboratory for Applications of Remote Sensing (LARS), the video tapes mentioned below have recently been added. None is available in alternative formats. Although the specific case example material relates to research projects developed by LARS staff, all of these programmes are appropriate to higher education students specialising in remote sensing. Titles include: Introduction to Quantitative Analysis; The Role of Pattern Recognition in Remote Sensing; Correction and Enhancement of Digital Image Data; The Role of Numerical Analysis in Forest Management; Multispectral Properties of Soils. (H) (30 minutes in each case, 1981). Each videotape is supported by 25 copies of the viewing notes and the text by P H Swain and S M Davis (1978) Remote Sensing: The Quantitative Approach, McGraw-Hill.

University of Arizona: have produced a set of 14 videotapes, for rental or purchase, on the theory of remote sensing and on image processing. The level of presentation is specific to senior undergraduates and postgraduates. Details of the series, which are available in a variety of formats can be obtained from: Microcampus, University of Arizona, Civil Engineering Building, Room 72, Tucson, Arizona 85721 USA (H).

Teachers in higher education should be aware of the facilities offered by the British Universities Film and Video Council Limited, 55 Greek Street, London W1V 5LR. The Council maintains a comprehensive information service and library relating to films and video materials produced worldwide. Reviews and the evaluations of users are recorded. Institutional or private membership is available and it offers privileged loan of films, attendance at conferences and workshops, etc at reduced cost and access to facilities. In the last two years the BUFVC has been active in promoting the use of film and video in the teaching of remote sensing.

Films and videos specific to photogrammetry, aerial survey and air photo interpretation are not included, but titles can be identified through the BUFVC. Their exclusion is justified because the majority are in excess of 15 years old and are no longer available for hire. There are also numerous films that employ oblique aerial views as the medium for illustration, especially within geography, environmental studies, earth science, archaeology and economic history.

C. POSTERS AND WALLCHARTS

This is a somewhat difficult entry to categorise, as some posters use image mosaics. These could be listed under image maps, but are reserved for this section where they lack any form of overprinted annotation and/or minimal geographical location detail.

1. UK National Remote Sensing Centre (NRSC)

(i) Britain and Ireland: Simulated natural colour mosaic of 52 Landsat scenes, 480 x 710 mm, £2.90.

(ii) Britain from 900 kms: Landsat false colour mosaic, 480 x 710 mm, £1.75.

(iii) Isle of Lewis: Colour coded Landsat image land classification, 660 x 810 mm, £3.45.

(iv) The Third Planet: True colour Meteosat image of full disc Earth, 400 x 610 mm, £1.75

The NRSC 'Fact Sheet' series consists of 24 A4-sized, colour-illustrated summaries of both general remote sensing topics and details of NRSC facilities and services. They constitute very attractive display material, but ideally teachers require two copies of each as there is reverse printing on single sheets. All Fact Sheets come in two series and are available free of charge directly from the NRSC: (1) Facilities and Services: User Services; Photographic Products; Special Products; Mobile Exhibition; Metsat Services; Applications; Image Processing; Interactive Image Analysis; (2) Satellites and Sensors: Landsat 1, 2 and 3; Landsat 4 and 5; Meteosat; NOAA Polar Orbiting Satellites; Seasat-Synthetic Aperture Radar; Heat Capacity Mapping Mission; Nimbus 7 Coastal Zone Colour Scanner; Shuttle Imaging Radar; Photography from Space; SPOT; ERS-1; Radarsat; and a number of other titles relating to applications fields in which the NRSC has made a direct contribution.

2. Remote Sensing Sales Centre and McCarta Limited

Accredited NRSC distributor, all of the above-listed posters, some of which are also available as high gloss prints, each at £7.50 (240 x 240 mm) or £14.30 (480 x 480 mm), can be obtained here. The Remote Sensing Sales Centre has been set up to supply all kinds of Remote Sensing derived products and carries an extensive stock of books, textbooks, atlases, image interpretation sourcebooks, manuals, computing resources, overhead transparencies, slide sets, photographic prints, posters, calendars, videos and satellite image maps. A catalogue is available and information about new products will be welcome. Clients can call at the McCarta Remote Sensing Sales Centre Shop, 122 King's Cross Road, London WC1X 9DS (01-278 8276), a mail and telephone order service is also available.

3. ERSAC Scientific Publications (ESP)

Photographic prints of Landsat false colour and simulated natural colour mosaics are marketed by ESP. Other ESP products include a set of six black and white postcards of weather satellite scenes of North-Western Europe.

4. Nigel Press; Clyde Surveys; Hunting Surveys

Both controlled and uncontrolled mosaics of Landsat imagery for a wide variety of different areas, including the British Isles, have been constructed for survey work, or can be composed at the request of individual users. Although in no sense posters or wallcharts, they are mentioned here because they are composed without marginal or superimposed locational or other detail. Users with specific geographical requirements will find the archives of all three companies of value. Hunting Surveys have produced two posters of Seasat SAR (image mosaics, one at 1:2,000,000 of the entire British Isles and another of Iceland at 1:500,000). Nigel Press publish an index to all available mosaics, at a variety of scales; their coverage of Africa is complete. Further details are available in the comprehensive Data Centre catalogue issued by Nigel Press Associates.

5. Department of the Environment

The DOE published a short series of wallcharts of Landsat colour scenes (and selected sub-scenes) of certain regional areas of England (the south east; Midlands; north west) in 1976. Each scene is accompanied by an explanation, in straightforward terms, of the sensor system and some of the identifiable detail. Reproduction quality is modest, but prices are low. These posters will not be reprinted, and are available only whilst remaining stocks last. Order (without accompanying payment) from: Department of the Environment, Map Intelligence Branch, Albert Embankment, London SE1 7TF (01-211 3000).

6. P and J Storey

Four black and white satellite maps at 1:200,000 of the Lake District; North Wales; Northern Pennines (Dales) and Southern Pennines (Peak) are marketed at £1.85 each by P and J Storey, 10 Rosebank Close, Cookham, Berks.

7. Space Frontiers Limited

Space Frontiers Limited market a range of charts and posters, mostly featuring images of the planets of the solar system as well as subjects such as rockets, space shuttle, etc. Most are originals from NASA, including a large format poster Comparing the Planets (56" x 30"). A set of full colour Astro Palomar Posters featuring Nebulae, Pleiades Star Cluster and the Andromeda Galaxy may be purchased individually. In addition two sets of postcards devoted to (i) astronomical objects and (ii) a selection of deep space mission images of the planets of the solar system are particularly attractive to schools. The same is true of a folio of 13 lithographic pictures of some of the more dramatic images returned from space missions between 1968 and 1979. The latter are ideal as background display material for school project work; explanatory captions are printed on the reverse side, but may be conveniently photocopied. The general title of this collection is Frontiers of Space.

Space Frontiers can only supply whilst existing stocks last. A catalogue of all products, and a price list, are available from the company at 30 Fifth Avenue, Denvilles, Havant, Hants, PO9 2PL (0705 475313).

8. Environmental Resources Analysis Ltd

ERA market a glossy natural simulated colour poster of a Landsat mosaic of Ireland, at a scale of 1: 3,000,000. It is available from ERA, 187 Pearse Street, Dublin 2 (0101-77 2922 ext 1244).

9. Other Suppliers

Spacecharts is a publishing company that is dedicated to the production of large format educational wall charts featuring integrations of imagery, text, diagrams and captions on a variety of space themes. The three so far available are: Space Shuttle, Saturn, and Jupiter at £2.00 each. The style of presentation and explanation makes them particularly suitable for upper secondary and further education schools and colleges, as well as museums etc. Further details from: Spacecharts, Newton Tony, Salisbury, Wilts, SP4 OHF (098 064 672).

Aerofilms Limited market an annual Flying Calendar consisting of 12 49 cm x 49 full colour photographs of aerial and satellite images, accompanied by full caption explanations. The selection is worldwide, and the individual pictures are effective wallcharts or posters. Back numbers may be available, thus providing a visually stimulating cross-section of a wide variety of image types. The price of the 1984 edition was £9.75. Details from Aerofilms Limited, Gate Studios, Station Road, BorehamWood, Herts, WD6 1EJ (01-207 0666).

Outside of the UK, wallcharts and posters are available from NASA (a very wide selection, mostly geared to secondary and primary school display) and Pilot Rock Inc. The latter market a space portrait of the United States, a 36" x 46" glossy finish poster of a Landsat colour mosaic of 569 cloud-free images of the conterminous USA. The reverse side reproduces 18 5" x 5" Landsat sub-scenes of selected US cities and areas of mineral and other resource exploitation and development. A poster, Canada from Space, is published by Space Dimensions, PO Box 3022, Station C, Ottawa, K1Y 4JS, Canada, in false colour. A simulated natural colour photomosaic of Landsat scenes, at a scale of 1:4,560,000, titled Portrait USA was published by the National Geographic Society, Washington DC, 20036, USA, in 1976.

The International Training Centre for Aerial Survey and Remote Sensing has published a small number of wallcharts specific to certain remote sensing technical processes and the nature of the electromagnetic spectrum. They consist of linked sets of annotated diagrams suitable for higher education and professional training purposes.

EARTHNET have recently released a colour poster of Landsat Thematic Mapper imagery of Rome and its regional area. Available from the publishers or via the NRSC.

PART 2.4: SATELLITE IMAGE MAPS

There is a growing, though still limited, worldwide collection of officially published maps based directly on satellite images. They are based on colour and black and white mosaics of selected imagery that have been geometrically corrected to match a selected map projection, and combined selectively with existing map information. Features such as major landforms, urban areas and vegetation patterns are reproduced directly from the imagery and are supplemented by printed information such as place-name annotation, transport networks, latitude and longitude graticules, etc.

The majority of these satellite maps have been based on either Landsat MSS or Landsat RBV imagery with a few employing Skylab originals. The majority have been published by the US Geological Survey and cover the conterminous USA. A selective listing of those in print (at the time of compilation of this book) appears below.

A. MAPS OF THE UNITED STATES

1. Non-Serial Map Editions

1. Arizona, Black and white mosaic and map based on same (two sheets), US Geological Survey, 1974 (W).

2. Satellite Mosaic of Florida, US Geological Survey, 1974 (E).

3. Land Cover Map of Coastal Plain of Arctic National Wildlife Refuge, Alaska, Sheet No 1-1443, 1983, 1:250,000 US Geological Survey (W).

4. State of New Mexico: Mosaic of Band 5 ERTS-1 Imagery ... Koosle and Pouls Engineering, 1973.

5. ERTS (Band 5) Orthophotomosaic Map of Missouri, Missouri Geological Survey 1974.

6. Grand Canyon National Park, 1975, (three relief diagrams and text), National Geographic Society, Washington DC, 1971.

7. Heart of the Grand Canyon, Landsat 1 and 2 mosaic with accompanying geological cross-section. US Geological Survey.

8. The Nation's Capital, 1972, colour composite of Landsat 1, US Geological Survey, (E).

9. Landsat Image of Eastern West Virginia and Landsat Image of Western West Virginia both produced by West Virginia Geological Survey, 1974.

10. Landsat Mosaic of Oregon, based exclusively on Landsat 3 RBV imagery and USGS topographic base map detail,

1983. Environmental Remote Sensing Application Laboratory (ERSAL), Oregon State University, Cornwallis, OR 97331-6703, USA.

2. Serial Map Editions (US Geological Survey)

All are based on Landsat MSS imagery, unless otherwise stated.

ALASKA				
Teshekpuk	Black/white RBV mosaic	1:250,000	1983	W
Beechey Point	Black/white MSS mosaic	1:250,000	1983	W
Chandler Lake	Black/white RBV mosaic	1:250,000	1983	W
Arctic : North Slope (25 sheets)	Black/white RBV mosaic	1:250,000	1983	W
ARIZONA	Sepia mosaic	1:500,000	1975	W
Phoenix	Sepia mosaic	1:250,000	1975	W
CONNECTICUT				
Hartford	Colour Photomap (Skylab)	1:250,000	1976	E
DISTRICT OF COLUMBIA				
Washington DC and vicinity	Colour image, enhanced (Skylab)	1:100,000	1984	E
FLORIDA				
Pensacola Bay	Colour mosaic	1:500,000	1975	E
Lake Seminole	Colour image	1:500,000	1977	E
Apalachee Bay	Colour image	1:500,000	1977	E
Okefenokee Swamp	Colour image	1:500,000	1977	E
Gulf Hammock	Colour image	1:500,000	1977	E
Lake George	Colour image	1:500,000	1977	E
Charlotte Harbour	Colour image	1:500,000	1977	E
Lake Okeechobee	Colour image	1:500,000	1977	E
Sanibel Island	Colour image	1:500,000	1977	E
The Everglades	Colour image	1:500,000	1977	
Florida Keys	Colour image, enhanced	1:500,000	1977	E
GEORGIA	Colour mosaic	1:500,000	1976	E
MARYLAND/VIRGINIA				
Chesapeake Bay and vicinity, Winter 1976-1977	Colour mosaic	1:500,000	1978	E
Upper Chesapeake Bay	Colour image, enhanced	1:250,000	1977	E
Upper Chesapeake Bay	Colour image, enhanced	1:500,000	1976	E
MASSACHUSETTS				
New Bedford	Black/white RBV mosaic	1:100,000	1983	E
MONTANA				
Pumpkin Creek	Colour image	1:500,000	1979	W
NEVADA				
Las Vegas	Colour mosaic with map on reverse	1:250,000	1983	W

NEW JERSEY	Colour mosaic	1:500,000	1983 E
TENNESSEE			
Dyersburg	Colour image/topo map on reverse (Landsat TM)	1:100,000	1983 E
UTAH			
Great Salt Lake vicinity	Colour, enhanced (Landsat TM)	1:125,000	1985 W
WASHINGTON			
Wenatchee	Colour mosaic with planimetric base	1:250,000	1979 W
WYOMING			
Medicine Bow River	Colour image	1:500,000	1979 W

The code W or E refers to the distribution source: W = Western Distribution Branch, USGS, Box 25286, Federal Center, Building 41, Denver, Colorado 80225, USA; E = Eastern Distribution Branch, USGS, 1200 South Eads Street, Arlington, Virginia 22202, USA.

Although designed as a basis for class exercises with junior forms (10-14 years), the Primer for Satellite Maps by J I Kirman is a pioneering publication that reproduces a variety of Landsat MSS colour composites - exclusively of Canada - that are used directly as maps. Published by Puckrin's Production House, 35 Mill Drive, St Albert, Alberta, Canada T8N 1JS in 1978, it costs C$ 40.00. Imagery reproduced in other atlas-style publications, interfaced with published or specially designed corresponding maps, also fulfils a similar role (see page 77).

A number of national mapping agencies and commercial companies have produced a variety of controlled satellite image mosaics without additional annotation or with only place names and a few marginal details added. These are listed below, on a selective basis:

1. Land Cover Maps of Washington DC Area published as A Folio of Land Use Maps of Washington DC by the US Geological Survey in 1978. These six maps depict land use determined by supervised computer classification of 1972 and 1973 Landsat MSS data. Map No. 1-858-E is overprinted with locational detail; Map No. 1-858-F has overprinting of enumeration district boundaries and 1970 Census Tract information. Land cover statistics are given on the reverse of each map. The cost is $11.75 for each complete folio; from: USGS Branch of Distribution, 1200 South Eads Street, Arlington, Virginia 22202, USA.

2. Five contact mosaics of the State of Alaska, at a scale of 1:1,000,000, and a single combined mosaic at 1:250,000 available in a variety of sizes and prices. They were published in 1978 and are available from: USGS, Eros Data Center, User Services, Sioux Falls, South Dakota 57198, USA.

3. A Map of Mexico City and Vicinity from Space was published by Pilot Rock Inc., California (Publication PRl 174) in 1981, it is 21"x 28" and is accompanied by a key and description.

4. A complete photomap of the 48 contiguous United States assembled by the US Department of Agriculture Soil Conservation Service from Landsat 1 images was produced in 1974. In addition to this single sheet composite photomap of the 48 states, the US is sub-divided into 6, 17, and 54 individual sheets at various scales. A similar mosaic and series was constructed of Alaska in co-operation with the Resource Planning Team of the Joint Federal State Land Use Commission for Alaska. The Alaska mosaic is of Landsat Band 7 only.

Grids, information and orders for the above can be obtained from: Aerial Photography Field Office, ASCS-USDA, 223 West 2300 South, PO 30010, Salt Lake City, Utah 84125, USA (these can also be obtained from McCarta Ltd).

5. A number of photomosaics have been compiled by the World Bank and are supplied by International Mapping Unlimited, 4343 - 39th Street, NW Washington DC, 20016, USA. There are:

1. Peru - Central 1:100,000, two sheets, 1982.
2. Bhutan - Soil, Water Resources 1:250,000, 1982.
3. Nepal - Land cover indications derived from Landsat imagery; colour, two sheets, 1:500,000, 1982.
4. Nepal - Land cover indications derived from Landsat imagery; two colour (brown and black), two sheets, 1:500,000, 1982.
5. Orissa, India - Land Cover and Land Use Association; 1:1,000,000, 1977.
6. Developing countries - Landsat Index Atlas of the Developing Countries of the World. 14 Maps; 1:10,000,000, 1976 (no later editions).
7. Bangladesh - Land cover, soil and water, 1:500,000, 1981.
8. Bangladesh - Land use classification related to land cover, 1:500,000, 1981.
9. Bangladesh - Land cover indications; two colour (brown and black), 1:500,000, 1981.
10. Burma - Land Cover and Land Use Association, 1:1,000,000, 1976.

B. MAPS OF OTHER COUNTRIES

1. Afghanistan

Satellite image map of Afghanistan, 1:2,000,000, Omaha 1977.

2. Antarctica

(i) Antarktis Satellitenbildkarte, 1:3,000,000, 1982
- West Neuschwassenland, 1981
- Neuschwassenland, 1982

(ii) Antarctica - USGS

1) Antarctica. Black/white Mosaic, 1:15,000,000, 1977
2) Ellsworth Mountains. Blue Tone Mosaic, 1:500,000, 1976
3) Victoria Land Coast. Blue Tone Mosaic, 1:100,000, 1976
4) McMurdo Sound Region. Black/white, 1:250,000, 1975
5) McMurdo Sound. Black/white, 1:500,000, 1975
6) McMurdo Sound Region. Blue Tone Image, 1:1,000,000, 1975

(iii) Antarctica - Division of National Mapping, Canberra, Australia: A series of Landsat photomaps at 1:500,000 and 1:250,000 covering the Australian Antarctic Territory (only available as dyelines).

(iv) British Antarctic Territory: photomaps (using Landsat) published by DOS (now part of Ordnance Survey).

3. Australia

South Australia Landsat imagery map at 1:500,000; 18 maps in total.

4. Austria

Osterreichische Satelliten Bildkarte, 1:200,000, 1983, Vienna.

5. Belgium

Multispectral Analysis of the Brussels Region. 1:37,000, two sheets, 1984, IGN Brussels. La Belgique: Vue de l'Espace (includes Luxembourg). 1:350,000, 1980, Belfotop-Eurosense PVBA, Vanmer Vekenstraat, 158, Wemmen, B-1810 Belgium.

6. Bolivia

Landsat-Fotomosaik Bolivien, 1:2,000,000, 1980. Carte Structurale des mes Septenthonaces de Bolivie Servisio Geologico de Bolivia, 1976.

7. Canada

These are produced by: Department of Energy, Mines & Resources,

615 Booth Street, Ottawa, Ontario, K1A DE9, Canada.

All of Canada	1:5,000,000	)
All of Canada	1:2,500,000	) all
Canada North of 60°	1:10,000,000	) black/
Canada North of 60°	1:5,000,000	) white
Canada North of 60°	1:2,500,000	)

Landsat Mosaics of Canada

Geometrically corrected photomosaics at a scale of 1:2,500,000 using black and white Band 6 (and a few Band 7) summer scenes. There is one additional sheet of Manitoba, using winter imagery. Sheet numbers and names:

1) Arctic Islands
2) Yukon
3) Great Bear
4) Thelon
5) Wager Ray
6) Baffin
7) Quebec North
8) St Lawrence
9) Ontario
10) Manitoba
11) Saskatchewan and Alberta
12) British Columbia

The above photomosaics were published in 1978. In 1980 two sheets at 1:1,000,000 for Newfoundland and the Maritime Provinces were added.

8. Central Europe

Satellitenbild-Mosaik Von Mitteleuropa 1:1,000,000, 1975; and Mosaic of Landsat Images, Bands 5 & 7, black/white, can be obtained from: Geocentre, Postfach 80 0830, 10-7000, Stuttgart 80, German Federal Republic.

Bundesrepublik Deutschland/DPR-Wectravmbildkarte 1:500,000, can be obtained from Westermann, PO Box 3320, Georg-Westermann Allee 66, D-3300, Braunschweig, German Federal Republic.

9. China

Geoscience Analysis of Landsat Imagery - An Image Atlas of China (1984) Chinese Edition (with English summary); an English edition is to follow (1986). 533 Landsat images 1:500,000, selected from Bands 4, 5 and 7 and derived from Landsats 1-3. These are published by Science Press, Beijing, China, on behalf of the Chinese Academy of Sciences.

10. Colombia

Landsat - Colombia, 1:2,000,000, 1980.

11. Egypt

Geological Map of Egypt Landsat imagery map 1:500,000, 1978: including the Aswan Quadrangle, Qena Quadrangle available from: Geological Survey of Egypt, Cairo.

12. Finland

Finland from Satellite 1:1,000,000, 1980, and The Nordic Countries from Satellite 1:2,000,000, 1981, both available from: Natural Resources Board of Survey, Helsinki, Finland.

13. France

La France Vue de Satellite (Infrared colour) IGM
1:1,300,000 £4.15

La France Vue de Satellite (Black and white) BRGM
1:1,000,000 £9.00

La France Vue de Satellite (Black and white) BRGM
1:2,000,000 £2.25

La Region Parisienne Vus de Satellite (Infrared colour) IGM
1:100,000 £4.95

Les Environs de Paris Vus de Satellite (Natural colour)
Simulated true colour IGM 1:100,000 £4.95

La Chaine des Alpes Vue de Satellite (Black and white)
Landsat 1 BRGM Orleans (France, east to Czechoslovakia)
1:1,000,000 £6.50

Series of Picto Images at 1:500,000 of France in natural simulated colour, published by BEICIP.

Alpes-Cotes D'Azur	£ 6.00
Corse	£ 5.50
Cote Atlantique	£ 6.50
Bretagne-Vendree	£ 6.50
Pyrenees-East	£ 6.50
Pyrenees-West	£ 6.50
Pyrenees-East and West (two sheets)	£11.00

14. Hungary

Satellite Image Map of Hungary 1:500,000, 1982: available from: Cartographia, Budapest.

15. Iceland

Vatnajokull colour image 1:500,000, 1976, USGS; and Vatnajokull Black/white (winter) 1:500,000, 1977, USGS. Both are Landsat imagery maps.

16. Indonesia

Sulawesi: Indonesia 1:1,000,000, Vancouver, 1977.
Peta Inders Indonesia 1:7,500,000, No. 1 Citra (Landsat).

17. Israel

Israel Geological Map with Satellite Imagery, 1:500,000, £11.75.

Sinai, Geological Satellite Photomap, 1:500,000, £15.35.
Satellite Photomap of Israel, 1:750,000, £7.50.
Department of Surveys, Tel Aviv, Israel.

18. Morocco

Le Maroc Vue de Satellite 1:1,000,000, two sheets. The following titles are available in both Arabic and French:

a) Nature
b) Zones Economiques et Emplacement des Gisements Minières en Surcharge.

Le Maroc Vue de Satellite 1:2,000,000, one sheet. The following titles are available in French only:

a) Nature
b) Avec le Code Structural en Surcharge

19. New Zealand

New Zealand from Landsat 1:2,000,000, New Zealand Department of Lands and Surveys, £4.25.

20. South Africa

There are a number of photogeological maps at 1:250,000 and 1:500,000 derived from Landsat, and photogeological maps at 1:100,000 for a large number of sites, available only as diazo prints. Maps and further information are available from: The Chief Director, Geological Survey, Private Bag, Z112, Pretoria 001, Republic of South Africa.

21. Spain

Mapa Fotos Peninsula Iberica 1:2,000,000. Also available at 1:1,000,000.

Mosaico Fotografico de la Peninsula Iberica E Islas Balearies Mappa de Lineamentos Deducidos de las Imagenes Landsat 1:1,000,000 £24.35.

Mapa Tematicol de Usos del Suelo de la Provincia de Madrid: Imagenes de Landsat 1:500,000, 1978. Both are available from: Instituto Geografico Nacional, Madrid.

22. Switzerland

Atlas of Switzerland - 16 Landsat Images of Switzerland 1:500,000, 1984 produced by: Kummerly and Frey, Berne.

23. Venezuela

Mosaicos de Imagenes de Satelites Landsat 1 y 2 de Venezuela 1:3,000,000, produced in 1977 by: Ministerio de Energia Y Geologia, Caracas.

24. West Germany

Satellitenbildkarte Bundesrepublik Deutschland 1:1,000,000,

1980 and Topographische Ubersichtskarte, Satellitenbildkarte 1:200,000, 1981 are both published by Institut fur Angewandte Geodagie, 6310 Frankfurt-Am-Main West.

Satellitenbildkarten von Medensachsen, Hamburg und Bremen 1:500,000, 1982.

Satellitenbildenkarten Schleswig-Holstein, Hamburg, Bremen and Angren 1:500,000, 1979.

25. Yemen

Yemen - geographic; colour, 1:500,000, 1978, USGS and Yemen - geological; colour, 1:500,000, 1979, USGS.

Uncontrolled mosaics of Landsat imagery are available, on request, from several commercial companies in the UK (see p. 109). Most lack any form of annotation. Coverage of Africa and the Middle East is particularly strong.

For information on poster mosaics of the British Isles, see page 108.

PART 2.5: MANUALS, CLASS EXERCISES AND OTHER TEACHING AIDS

A. RESOURCES FOR TEACHERS

Laboratory Manual for the Study of Remote Sensing, K. Lee, Colorado School of Mines (Golden, Colorado, 80401, USA), 1976, 255 pp. $6.00.

Remote Sensing Laboratory Manual, F. F. Sabins, 1981, $15.00, with accompanying Instructor's Key for Remote Sensing Laboratory Manual, $7.00. A comprehensive compendium of interpretational and computational exercises suitable for tertiary level students. The main sub-divisions are: Aerial Photography, Manned Spacecraft Imagery; Landsat Imagery; Thermal Infra-red; Radar; Digital Image Processing; Resource Exploration (minerals and groundwater); Land Use and other environmental applications, and Comparison of Image Types. Within the systematic divisions, geological, geomorphological and oceanographical applications dominate. Despite this bias, and the fact that most of the worked examples are from North America, each exercise is self-contained and backed by adequate background descriptive detail. All of the exercises are supported by a set of 100 slides which reproduce all of the original maps, diagrams and images in the Manual - available for $125.00. The two complementary resources enable students to work semi-independently; the Instructor's Key gives correct answers to computation questions, completed map exercises and preferred responses to interpretational questions. The slides are also accompanied by full descriptive notes which give the user information on the salient characteristics of areas covered by imagery. Remote Sensing Enterprises also markets various other slide sets that relate to the Manual sections noted above. These are listed in more detail on p. 100. Teachers will find this Manual a source of numerous ready-made exercises although they may want to obtain class sets of the originals, via the publishers. At the very least, the Manual provides a source of ideas for the construction of interpretational exercises using image examples more appropriate to specific student groups. The book is available from: Remote Sensing Enterprises, PO Box 2893, La Habra, California 90681, USA.

Laboratory Manual to Remote Sensing of the Environment, B. F. Richason (see p. 66).

Everyone's Space Handbook D. Kroeck, (1976), 176 pp., $9.95, is a manual listing sources of aerial photography and satellite imagery in the United States, compiled with needs of teachers and lecturers in mind. It is unfortunate that this extremely helpful guide is now out of date in several respects, having been overtaken by institutional changes as well as the data base itself. There remains a good deal of pertinent content, especially with reference to older material and air photo archives in the USA. The book also presents a digest of basic concepts of remote sensing, sensor systems, image products and scale/format factors, and provides a basic primer in remote sensing for teachers without any previous background. The publisher also markets a large collection of remote sensing slides, which are listed on p. 99. For further details contact: Pilot Rock Inc., PO Box AS, Trinidad, California 95570, USA.

The National Aeronautics and Space Administration (NASA) provides a wealth of publications, resource materials, guides and directories designed specifically for teachers at all levels. Unfortunately, many of these are not available on request to non-US enquirers, but it is worth asking for current lists of publications under headings such as 'Earth Science and Environmental Education' from: Superintendent of Documents, US Government Printing Office, Washington D.C., 20402, USA. Information on new publications and materials will be sent to those who request to be placed on the 'Educators Mailing List' maintained by NASA Headquarters, Washington D.C. 20546. (Requests from overseas addressed to any of the regional centres of NASA are unlikely to produce a response). A leaflet on 'NASA Education Services' is available. For a relatively small outlay, teachers can obtain a variety of image examples, posters, fact sheets, graphics, brochures and background papers for class use. Most products are suitable for the 10-18 age range. For teachers in London, copies of many of the products can be previewed via the Educational and Cultural service of the US Embassy. As NASA has responsibility for the space programme in general, a large number of publications are marginal to remote sensing. It is therefore advisable to specify an interest in this field of space science and technology in order to obtain relevant information. Very few of the publications are available free of charge, but few are more than £2-£3 each.

Publications and other products arising from NASA's deep space planetary missions, for example the continuing Voyager programme, are available from the Jet Propulsion Laboratory, California Institute of Technology, 4800 Oak Grove Drive, Pasadena, CA 91103, USA. For teachers wishing to present the results of these remarkable extra-terrestrial remote sensing surveys, with all their inherent exhilaration and originality, this is the primary source of advice on resource availability.

Satpack I and Satpack II are two complementary teaching packs (although each can be used independently) designed to provide background material and classroom exercises on meteorological satellite imagery and its interpretation. In each case, selected imagery is linked with sets of data relating to nearly simultaneous ground and upper air measurements. A full description of the general weather situation, and a completed rephanalysis for a single day is given, followed by imagery and ancillary data for the next day. The latter is the subject of a set of questions that test observational and analytical abilities, using the worked example as a model. In Satpack I, a rapidly changing situation over the NE Atlantic and Western Europe in October 1979 is used; in Satpack II, changes apparent between 13 and 14 May 1979, over the same area, are exploited. Detailed teacher's notes accompany each pack. This unique resource was developed by staff at the Department of Meteorology, University of Reading, under the auspices of the Royal Meteorological Society, with students of A-level geography and O-level meteorology in mind. However, it has been widely adopted in the teaching of atmospheric science by higher education establishments. The packs are available for sale at £3.80 from Focal Point Audio Visual Ltd, 251 Copnor Road, Portsmouth, PO3 5EE, and from the Royal Meteorological Society, who supported its production.

The Applications branch of EROS Data Center has published a long set of 55 Workshop Exercises specific to the application of remote sensing data to natural resource inventory and management. They are the product of the Center's experience in providing professional training courses, and each consists of a set of photographic copies of images, relevant overlays and exercise instructions. Exercise solutions and model answers are also provided. They are adaptable to independent learning, but are designed to be complementary to formal lectures, thus representing a 'ready made' set of practical exercises. Most are essentially introductory in level and do not need to be supplemented by discussion and additional demonstration and teaching. Whilst some of the exercises are of a general nature (e.g. 'Introduction to Multispectral Characteristics of Landsat Imagery'; and 'Measurement of ... Plant Parameters and Vegetation Change Using Very Large-Scale Color Infrared Photographs') others are specific to one application example. The latter are all based on US projects, e.g. 'Regional Soils Mapping in South-eastern South Dakota using Landsat Imagery' and 'Targeting Ground-Water Exploration in South Central Arizona...'. Others are intermediate between these, such as 'Delineating Flood Boundaries on Aerial Photographs' and 'Urban Land-Use Mapping in a Semi-Arid Region using Landsat Imagery'. Prices vary between $35 and $100, according to the number of photographs and other accompanying materials. Film transparencies of the original photographs may also be purchased for the production of large class sets. Full details can be obtained from: User Services, EROS Data Center, US Geological Survey, Sioux Falls, South Dakota, 57198, USA. EROS also distributes poster and reference material of relevance produced by the USGS National Mapping Division. An example of this is a large poster format index of data from major US satellites (1985, free on request).

The Landsat Tutorial Workbook. NASA, Reference Publication 1078, 1982 is a large format, extensively illustrated set of lecture notes with specific guidance and recommendations on practical exercises (though no practical exercises as such are included). This resource is adaptable to various educational levels. Available from the USGPO, Washington D.C., 20402.

Pilot Rock Inc. (see p. 124) also publish a small number of Teaching Modules, two of which are attractive for use in schools and colleges. That on The San Francisco Bay Region from Space includes a large lithographic print (in colour) of a high altitude aerial photo, an accompanying map, student laboratory exercises and key, and an instructor's handbook (including lecture notes and overlay diagrams). The pack costs $21.50 (1981). The module on Mount St Helens includes a set of posters and a sequence of 40 slides of both remotely sensed and ground views of the area before, during and after the eruption of 18 May 1980. The slides are accompanied by detailed lecture notes, which include suggestions for practical work. The attraction of the slide set, quite apart from the compelling nature of the subject, is the use of a wide variety of different image types and scales, e.g. Landsat MSS image (before eruption); NOAA-6 imagery of the scene a few minutes after the major blast and the pattern of ash fall-out over the next five days; high altitude colour infrared views of the mountain before and after the eruption event, and infrared thermal imagery, colour coded to show temperature gradients, during the monitoring period prior to the eruption. Two books, Fire and Ice: The Cascade Volcanoes, and Volcano - The Eruption of Mt St Helens and a map with numerous topographical and geological sections may also be purchased as supplementary materials. The complete collection costs $64.00, but individual components may be purchased separately.

Aerofilms market a Remote Sensing Set, Mark II (which replaces an earlier version). It is primarily designed for use in higher education classes, and contains a total of 46 prints of different types of imagery and accompanying extensive explanatory lecturing or teaching notes. The design of specific practical exercises based on this collection is left to users, though the notes contain several indirect suggestions. Most images are provided as stereo pairs, and four of the illustrations have interpretative overlays. It is a well-selected, balanced collection (though inevitably already overtaken by the development of new satellite imaging systems since its publication in 1981). It provides a ready-made set of imagery which would be difficult for individual teachers to assemble independently. The geographical location of imagery is worldwide, with most of the Landsat examples taken from the Middle-East. A descriptive leaflet of this resource, with current prices, is available from the publishers (a set of slides of all the images is also available, but cannot be purchased separately): Aerofilms, Gate Studios, Station Road, Borehamwood, Herts, WD6 1EJ.

The Atlas for Interpretation of Multispectral Aerospace Photographs, Moscow and East Berlin (1982), described on p. 79, is an excellent source book for large format, poster-style imagery suitable for classroom use.

B. OTHER TEACHING RESOURCES

1. Sets of Imagery for Class Use

The companies listed below market specially selected sets of imagery for class use, which may be purchased in multiple units or individually.

1. Aerofilms Ltd produce over 200 stereo-pairs of air photos under headings such as 'Geological Structure'; 'Coastal Scenery'; 'Rivers'; 'Agriculture'; 'Mining and Quarrying'; 'Industry'; 'Forestry'; 'Communications'; 'Towns and Villages', and 'Features of Historical and Archaeological Interest'. These have been selected with the requirements of teachers of geography and environmental science/studies in mind and are supplied with descriptive notes. Most are black and white, with a few in true colour. The vast majority of subjects are taken from the landscape of the British Isles. Both oblique and vertical stereo pairs are

available, though the former is a much more limited range (all obliques are in true colour).

Under a similar set of headings, Aerofilms also publish a number of 'Educational Sets' of oblique air photos of the British Landscape, which are particularly appropriate to O and A level geography classes. They are also available as high quality overhead projector transparencies. A range of simple stereo viewers, lightweight and folding into plastic wallets, are marketed for class or field viewing of vertical air photos in three dimensions.

A selection of some of these obliques is presented in The Aerofilms Book of England from the Air (Ed. S. Booth), Blandford Press, 112 pp. 1979, £5.75, and The Aerofilms Book of Scotland from the Air, (Ed. J. Campbell), Wiedenfeld and Nicholson, 1984, 160 pp, £12.95; available directly from Aerofilms. They are visually stimulating books that deserve a place in any personal or public library. The full selection of obliques used in the 'Educational Sets' series is published in The Aerofilms Book of Aerial Photographs, and they are also reproduced (at a small scale) in the company's publicity brochure.

Most of the subjects of the oblique photographs (but many others in addition) are available as sets of 35-mm slides. The comprehensive library and reference system of Aerofilms is detailed on p. 42. The company also market pocket and mirror stereoscopes.

The publicity leaflets and brochures setting out details of these products are exceptionally attractive and informative. These, and current prices, are obtainable from: Aerofilms Ltd, Elstree Way, Borehamwood, Herts, WD6 1SB (01-207 0666).

2. Clyde Surveys market an 'Educational Series' of vertical air photos according to topic, viz 'Geological Features'; 'Geomorphological Processes'; 'Agricultural Land Use'; 'Thermal Infrared Linescan'; 'TIR Linescan and Simultaneous Colour Photography' and 'Simultaneous Colour and Colour Infrared Photography'. The latter is divided into five separate subsets, devoted to coastal and inter-tidal vegetation and landforms; agricultural crop classification; conifer plantation mortality; vegetation stress analysis and urban land-use mapping. Additionally, there are two shorter sets on colour infrared mapping (specific to the recording of conifer tree types) and multispectral photography. The number of prints in each collection varies considerably. All are accompanied by descriptive and explanatory notes for teachers and students. The locations of subjects are worldwide, except for the sets on thermal linescan and multispectral photography which are restricted to England and Wales. These teaching sets are of particular value to teachers in higher education and provide good opportunities for exercises on the comparative attributes of different types of imagery from airborne sensors. (The company also markets a stereo viewer, a low cost stereoscope mounted on a baseboard for the three-

dimensional viewing of overlapping imagery.) Further details in the form of descriptive sheets of each set, and prices can be obtained from: Clyde Surveys Ltd, Reform Road, Maidenhead, Berks, SL6 8EU.

2. UOSAT

In a quite different category, and of special interest to science teachers, is UOSAT I and II, low cost 'educational' satellites developed and built by the University of Surrey, and launched in 1981 and 1984. They carry on board a number of experiments providing information on the Earth's magnetic field, solar activity and auroral phenomena. It is also equipped with an imaging system that can obtain 2km resolution pictures of 500 x 500-km areas of the Earth's surface, with enhancement of land features and land/sea boundaries. All forms of data can be transmitted directly to users with relatively inexpensive TV and radio receiver equipment (less than £300) to form the basis of project work in subjects such as general science, mathematics, physics and geography. Processed telemetry and experimental data can be displayed in graphical form, with telemetry also available in a digitally synthesised speech format. Schools and colleges with an interest in receiving 'live' data in this way should contact UOSAT at the Department of Electronic and Electrical Engineering, University of Surrey, Guildford, GU2 5XH (0483 71281 ext 755) for further details of the system, hardware requirements and applications potential.

3. The British Interplanetary Society

The Society, with a membership of over 3500, vigorously promotes interest in space science for pupils, students and teachers (as well as a wide spectrum of professionals and amateurs). Conventional remote sensing is only one of a wide range of topics that form the Society's terms of reference, but technical meetings, general lectures, film shows and visits, as well as a film hire service, are often of direct relevance to the subject. The Society's magazine, Space Education, is written specifically for pupils and teachers, and is available on subscription independent of membership. Full details are given on p. 159. Institutional membership (e.g. on behalf of schools, colleges, etc) is not available.

4. The Air Education and Recreational Organisation

An organisation for schools and colleges dedicated to the advancement of all aspects of aviation is The Air Education and Recreational Organisation. It has occasionally organised excursions concerned with 'geography from the air' and talks on the use and acquisition of aerial photographs. These provide a substitute for courses of a similar kind previously organised by the Department of Education and Science. Details of membership can be obtained from: The Secretary, Carwarden House, 118 Upper Chobham Road, Camberley, Surrey.

5. Museum Displays

The only permanent museum displays on remote sensing in the UK are at the Science Museum (South Kensington, London), the National Museum of Photography at Bradford and the Manchester Space Centre. The Geological Museum also has an exhibit on remote sensing missions to the planets of the solar system and the Moon. The bookshop of the latter stocks a few of the slide sets and posters listed earlier, which may be purchased 'over the counter'. Both the Science and Geological Museums sell a booklet titled Remote Sensing in their Exploring Science series.

6. Other

Numerous journal and magazine articles on the utilisation of aerial and space imagery in coursework at various levels have been published, of which the proceedings of the Conferences of Remote Sensing Educators (CORSE) 1978; 1981 are noteworthy (NASA). The latter are also concerned with the conceptual and practical problems of teaching remote sensing; contributors are exclusively from the USA and Canada.

7. Other Sources of Teaching Aids

There are no suppliers of teaching sets of satellite imagery (except for those examples in Aerofilms' Remote Sensing Set, Mark II, and the Royal Meteorological Society's Satpack.) However, Nigel Press Associates advertise teaching packs, each of ten Landsat images, to illustate (a) different wavebands and scales, and (b) certain geological and geomorphological features and ecosystem characteristics. These packs can be 'customised' to individual interests to some extent. Further details from: Nigel Press Associates, Old Station Yard, Marlpit Hill, Edenbridge, Kent, TN8 5AU (0732 865023).

A resource that is unfortunately no longer in print, but which is held by many schools and some local authority resource or teachers' centres is Air Photo Pack No. 1: A Selection of English Cultural Landscapes, compiled by the Geographical Association and published by C F Casella Ltd, Regent House, Brittania Walk, London, N1 (01-253 8581). It includes eight stereo pairs of photographic prints (9" x 9"), a transparency overlay with coded tracing of one photograph in each pair, a pocket stereoscope, air photo interpreter's transparent plastic scale and a booklet containing explanatory descriptions and suggested exercises for each stereo pair. Solutions and model responses to questions are included as teacher's notes. With one exception, all locations of the selected photos are from Greater London and Southern England. C F Casella also market a range of folding mirror and 'pocket' stereoscopes, detailed in their brochure No. 929/1985.

One of the authors of the above 'Pack', A D Walton, contributed A Topical List of Vertical Photographs in the National Air-Photo Libraries as No. 1 in the Geographical Association's Teaching Geography series (1967). It is a list of vertical air photos (held in the centralised regional libraries of the UK) ıat have been selected for their value for basic or more .dvanced geographical interpretation exercises. They are grouped under systematic headings such as landforms, woodland, communications, industry, etc. In each case precise identifying details are given (scale, date, sortie and print numbers). All of these photos should still be available by direct ordering from the relevant libraries listed under sources of aerial photography (see pp. 39-42). The paper may be obtained from The Geographical Association, 343 Fulwood Road, Sheffield, S10 3BP.

C. COMPUTER APPLICATIONS

Educational usage of digital image processing is limited by the high cost of purchase, operation and maintenance, although some universities and polytechnics with specialist remote sensing courses have acquired these systems. Charges for hiring time are also relatively expensive, though cost-effective for commercial users. Higher education establishments may be able to justify time on the GEMS or LS-10 systems installed at the NRSC, which can be arranged by appointment. The latter is particularly appropriate for teachers and students without previous experience of image processing, but both are essentially 'user-friendly'.

Another useful product is Weather Radar and Satellite Images on a BBC Micro, a package of 30 images illustrative of synoptic weather situations that are frequently encountered over the British Isles. It comprises two 80 track (or four 40 track) discs, accompanying software and a teaching manual. Produced in collaboration with the Meteorological Office, this package was published in 1985 by Computer Solutions for Science and Business, 73 Church Street, Malvern, Worcestershire, WR14 2AE (06845 65394), at £39.95.

Britain From Space: An Atlas of Landsat Images, and projected workbook and slides will be extended by the production of microcomputer software that will allow simple but instructive processing of Landsat sub-scenes. It is designed for use in schools, field study centres and other secondary education institutions. Provisionally titled the Landsat Image Processing Package, it is in active development at the time of writing. It is intended that schools will obtain sub-scenes of specific interest or limited to a defined area, e.g. the school catchment, a field study region or a particular case study that is the subject of a project investigation. Tentative publication is mid-1986; price not yet known. It is designed as flexible teaching and learning material for geography, geology, biology, physics and computer studies classes, and could be an exemplifying resource for courses in information technology. Enquiries should be addressed to the Remote Sensing Sales Centre (see p. 109).

Of interest and use to those with some previous experience of handling digital data on fairly large computers with FORTRAN programs is A. P. Cracknell (1982) Computer Programs for Image Processing of Remote Sensing Data, produced by the Carnegie Laboratory of Physics, University of Dundee, SS1 4HN, £10.00.

The 26 programs offer scope for the setting up of practical exercises with advanced students, but they do not represent a suite of maintained software.

As a final point, it is worth mentioning that some of the organisations listed throughout this book may be willing to accommodate visits from small groups of students. The NRSC at Farnborough has an Education Officer, who should be contacted in the first instance. Regional exhibitions are organised by the NRSC, NERC and other bodies, and are usually publicised in the educational press. Organisers of major conferences, workshop meetings, etc, might also enquire about a possible visit from the NRSC Road Show, a mobile exhibition facility equipped to demonstrate satellite imagery, concepts of image processing and application fields. It has image processing facilities,video programmes and facilities for lectures and discussions for small groups. It is managed by the Publicity and Promotions section of the Department of Trade and Industry, Millbank Tower, London SW1P 4QU, to whom initial enquiries should be addressed.

PART 2.6: EQUIPMENT, CONSULTANCY AND PROFESSIONAL SERVICES

The rapid development of remote sensing technology and analytical methods in the past 15 years has provided opportunities and markets for manufacturers of both simple and sophisticated equipment. There has also been a growth of consultancy organisations offering expertise in data analysis, image interpretation and application fields. In both cases, companies have developed markets and secured contracts overseas - often to the extent of dominating over their activities in the UK. The consultancy units of certain academic institutions have also attracted work from foreign sources, thus giving a valuable international dimension to remote sensing in this country.

Whilst some of the companies and organisations have been active for several decades, the majority are relatively new (or, at least, are new to remote sensing). Many of the entries in this section have been taken directly (with permission) from the brochure titled Remote Sensing, published by the DTI in November 1984. An asterisk (*) indicates that further details are given elsewhere in this book.

Full details of company, institution, etc, activities, equipment resources, personnel and product range are available in the DTI'S compendium on UK activity in remote sensing, Remote Sensing of Earth Resources (4th edition, 1981). This is now inevitably out of date as well as out of print. A fully revised Directory of Remote Sensing Activity in the UK is now in preparation. The Remote Sensing Yearbook will be a further source of updating (Volume 1 is due for publication in late 1985, by Taylor and Francis, Basingstoke and London). Many UK companies, agencies and institutions are corporate members of the Remote Sensing Society and contribute news on new developments to the Society's News and Letters.

A. EQUIPMENT AND PRODUCT MANUFACTURERS: COMMERCIAL CONSULTANCY COMPANIES

X

1. Biospec Ltd

Biospec Ltd is a research and environmental science consultancy specialising in the plant sciences, eg geobotany, biophysics, biochemistry, forestry and ecology, with particular reference to optical instrumentation, spectroscopic research and remote sensing. Staff provide research and feasibility studies, as well as commissioned surveys, in projects such as vegetation assessment for mineral deposits, crop condition, fertiliser requirement, agricultural and forest inventories and product estimation. Biospec Ltd, 18, Woodstock Road, London W4 1VE (Tel: 01-785 6341).

2. *Cartosat Production Ltd

Cartosat Production Ltd specialises in the use of satellite imagery to produce geometrically precise base maps for a wide range of different applications. Cartosat has developed new techniques in image correction and precision reprographics to achieve this end. Projects have been undertaken in North and South America, Europe, Africa, the Middle and Far East, and have involved both onshore and offshore work. The scale of these projects has ranged from small areas studied in great detail to mosaic studies of whole continents. Clients include United Nations organisations, government agencies and various industrial and private concerns. Cartosat Production Limited, Lansing Building, Old Station Yard, Marlpit Hill, Edenbridge, Kent, TN8 5AU (Tel: 0732 865023).

3. *Clyde Surveys Ltd

Clyde Surveys Ltd is an aerial survey and mapping company, formerly known as Fairey Surveys, using its own aircraft and in-house facilities to provide a wide range of products to its worldwide clients. The Environment and Resources Consultancy Division integrates the use of remote sensing techniques within the overall range of activities undertaken by the Clyde Group. This includes aerial photography, satellite imagery, thermal infra-red linescan, multispectral imagery etc. The acquisition and interpretation of this data produces natural resource inventories and monitors environmental change on a global, regional or project basis. Clyde Surveys Limited, Reform Road, Maidenhead, Berkshire SL6 8BU (0628 21371).

4. CW Controls Ltd

CW Controls Ltd is a firm of industrial engineers producing a range of special purpose electronic equipment for monitoring, control and test applications. It has developed a low-cost image processing system, the LS-10, (installed at the NRSC). CW Controls Limited, Industrial Electronics Engineers, Jubilee Works, Crowland Street, Southport, Merseyside PR9 7RR (70704 40623).

5. DIAD Systems Ltd

DIAD Systems Ltd specialises in development of turnkey installations for satellite image processing and geographic information systems. The company has written a very advanced and comprehensive suite of software in a high level language which gives scope to handle future data such as SPOT and TM imagery.

DIAD turnkey systems range from a low-cost micro-system based on the 68000 chip capable of handling data on the new EROS floppy disk format, up to powerful systems running on 32 bit super-minis such as the DEC VAX 11/780. A particular feature of the DIAD software is that it is 'user-friendly' and easy to integrate with other existing software.

DIAD Systems also undertakes contract research and development and is currently working on cartographic correction systems using satellite data and new geometric concepts to handle SPOT and other image data. DIAD Systems Limited, Lansing Building, Old Station Yard, Marlpit Hill, Edenbridge, Kent TN8 5AU (0732 865023). DIAD is a constituent of Nigel Press Associates.

6. *Environmental Remote Sensing Applications Consultants (ERSAC) Ltd

ERSAC Ltd has established a Remote Sensing Applications Centre in Scotland. This Scottish space centre is applications-oriented and dedicated to national and international promotion of remote sensing and digital mapping methods for natural resource surveys, environmental monitoring and hydrographic mapping.

ERSAC offers a geoscientific survey and mapping consultancy service using the GEMS image processing system. ERSAC has established a special data exchange facility to create a comprehensive worldwide satellite data tape archive.

This new centre also provides an international service to survey companies, resources based industry, government planning departments, research organisations and other agencies who require special image processing and interpretation facilities. Ersac Limited, Peel House, Ladywell, Livingston, Scotland EH54 6AG (0506 412000).

7. *Feedback Instruments Ltd

Feedback Instruments designs and manufactures a wide range of educational and test instruments for colleges, universities and industry. It exports two-thirds of its manufactured output through its seventy overseas agents, and has established a reputation for innovation and the quality of its products. Its latest development, the WSR513 weather satellite receiver is a complete, but low cost earth station capable of receiving weather pictures from polar orbiting satellites and, with the WSR 515 S-band adaptor, from geostationary satellites. Feedback Instruments Limited, Park Road, Crowborough, Sussex TN6 2QR (08926 3322).

8. Gems of Cambridge Ltd

Gems of Cambridge was formed in early 1983 to market the highly successful image processing system - GEMSTONE - which was developed by the Computer Aided Design Centre. Activities involve systems design, and manufacture and software development for a wide range of image processing applications, including digitization of photographs. The company have recently announced the availability of an advanced CCD scanning system (GEMSCAN) for very high resolute image analysis, and the GEMSTONE HANDBOOK (£16.00). The latter publication is a comprehensive introduction and manual for users. Gems of Cambridge Limited, Carlyle House, Carlyle Road, Cambridge CB4 3DN (0223 323818).

9. *Geosurvey International Ltd

Geosurvey International is a survey and exploration company with its own fleet of survey aircraft, including high altitude jets, undertaking aerial photography, airborne and ground geophysics and photogrammetric mapping.

The company has developed its own interactive digital image processing system to handle Landsat imagery and geophysical data. Interpretation of satellite imagery and aerial photography is performed for projects in mineral and hydrocarbon exploration, hydrology, engineering geology and agricultural development. Rectified satellite image maps are produced by techniques providing uniform map series for regional resource investigations. Geosurvey International Limited, Geosurvey House, Orchard Lane, East Molesey, Surrey KT8 OBY (01-398 8371/2, 01-397 0591/2).

10. Gresham Lion (PPL) Ltd

Gresham Lion (PPL) Ltd produces a comprehensive range of graphics and image display processors including high resolution systems for remote sensing applications including meteorology and earth resources. These are supported by the company's own specialist software packages for image coding, manipulation and enhancement, including a full Landsat package. Gresham Lion (PPL) Limited, Lower Way, Thatcham, Berkshire RG13 4RE (0635 62229).

11. *Hunting Surveys and Consultants Ltd

Hunting Surveys and Consultants is a multi-disciplinary survey company actively involved in remote sensing through its three operating divisions. The group claims more collective experience in the application of remote sensing techniques to resource surveys than any other in the UK.

The divisional companies have a great variety of equipment which includes: aircraft, airborne thematic mapper; 11 channel multi-spectral scanner, Hunting image processing and analysis system for interactive image analysis and preparation of enhanced hardcopy products; RC8 and RC10 air survey cameras; multi-spectral camera, magnetometers and gamma-ray spectrometers, colour additive viewer, geographic database system, digital mapping system linked to stereoplotters, geophysical data processing and plotting system, photographic laboratories and a library of aerial photography, satellite and radar imagery.

The three companies have 550 employees of whom 50 per cent are experienced graduate scientists. Many have specialised in the application, processing and interpretation of aerial photography, airborne and satellite radar imagery and digital processing and interactive analysis of airborne and satellite multi-spectral linescan imagery. Hunting also undertakes research into remote sensing applications and has been involved in the Seasat, Shuttle Imaging Radar (SIR-A), and SPOT simulation experiments.

Operating Divisions

Hunting Geology and Geophysics Ltd: Exploration for petroleum and minerals, airborne and ground geophysical surveys, processing and interpretation; airborne and satellite remote sensing and image analysis; geological mapping and deposit evaluation.

Hunting Technical Services Ltd: Agricultural planning, regional planning, environmental studies, soil and groundwater surveys, forestry and land use planning.

Hunting Surveys Ltd: Air photography, land, marine and engineering surveys, oceanography, photogrammetry, cartography, computer services, reprographic and photographic services. Hunting Surveys and Consultants Limited, Elstree Way, Borehamwood, Herts WD6 1SB (01-953 6161).

12. Joyce-Loebl

Joyce-Loebl manufactures image analysis instrumentation, exporting 75 per cent of its annual production. The company's products are used by medical, industrial and research scientists to undertake the detailed measurement and statistical interpretation of images such as specimen slides, satellite scenes and video film.

In 1981, Joyce-Loebl launched its latest television image analysis system, the MAGISCAN, which is claimed to be the smallest and most powerful image analysis system of its type.

All the major instruments are offered with complete interface and application software packages. Joyce-Loebl, Marquisway Team Valley, Gateshead, Tyne and Wear NE11 0QW (0632 822111).

13. Logica Ltd

Logica Ltd, is a leading computer and electronics systems and software company. Independent of both suppliers and users, the company carries out consultancy, design and implementation of meteorological and remote sensing systems and applications. Current activities include design of processing facilities for the ERS-1 satellite, development of synthetic aperture radar (SAR) processing and simulations facilities, design of onboard image compression systems and consultancy in infra-red sensor and processing technology. Logica Limited, Cobham Park, Downside Road, Cobham, Surrey KT1 3LX (09326 7355).

14. Marconi Research Centre Ltd

The Remote Sensing Group at Marconi Research Centre is contributing to the development of active microwave systems and applications, and the processing of microwave data. It has considerable experience in the design of synthetic aperture radar systems, digital processing, and image analysis and processing of radar data, as well as research into target interactions. It is also experienced in the design of scatterometer systems. It is involved with its sister company, Marconi Space and Defence Systems Ltd, (at Portsmouth) in the development of the microwave instruments for the ERS-1 satellite and in the development of advance radar processing facilities for the Royal Aircraft Establishment. Marconi Research Centre, West Hanningfield Road, Chelmsford, Essex CM2 8HN (0245 73331).

15. *Nigel Press Associates Ltd

Nigel Press Associates Ltd has specialised in the field of remote sensing since the launch of the first Landsat in 1972, and has accumulated many man years of experience interpreting and handling imagery from all over the world. Its Data Centre has one of the world's largest independent collections of imagery, with full facilities for digital image analysis, enhancement, display and plotting as well as photographic reproduction.

There is a staff of professional environmental and earth scientists trained and qualified in remote sensing which has carried out projects on the ground and with satellite imagery in five different continents. The range of projects covers many different disciplines, and further services are available through a wide range of associates. NPA is particularly active in geological work, and has undertaken tectonic studies on a continent-wide scale in Europe, Africa and the Middle East. Nigel Press Associates Limited, Lansing Building, Old Station Yard, Marlpit Hill, Edenbridge, Kent TN8 5AU (0732 865023).

16. Programmed Neuro Cybernetics (UK) Ltd

PNC (UK) Ltd combines the twin functions of acquisition and analysis of remote sensing data using both satellite and airborne data gathering equipment.

It has its own fully instrumented aircraft carrying unique multichannel infra-red scanners. PNC also has its own interactive image analysis and display system. Programmed Neuro Cybernetics (UK) Limited, Clivia House, 65 Old Church Street, London SW3 5BS (01-351 4311).

17. Sigma Electronic Systems Ltd

Sigma Electronic Systems Ltd is a company specialising in the development and manufacture of high performance computer graphics and image display systems for remote sensing and other applications. The firm has subsidiaries in France, Germany and the Benelux countries, and distributors in most other western European countries as well as Australia, the Far East and North America.

In addition to manufacturing a standard range of high resolution intelligent graphics and image display products, the company has a special systems group. This is an experienced team of systems analysts and hardware and software designers able to undertake the development of systems for individual customer requirements including turnkey information management systems based on satellite image data. Image processing systems have been supplied to government, military and academic research organisations in many countries. Sigmex Limited, Sigma House, North Heath Lane, Horsham, West Sussex RH12 4UZ (0403 50445).

18. Spectrascan Ltd

Spectrascan Ltd was formed in 1978, primarily to undertake the design and manufacture of systems and instruments to meet the needs of the remote sensing community. The staff also carries out consultancy work relating to remote sensing requirements and project viability. The company specialises in systems which combine optical, mechanical, electronic and computer technologies. Current remote sensing products include a family of spectrometers (spectroradiometer and MSS radiometer systems) and a range of laser film writers, complete with a full colour version, and an airborne MSS.

Other products include data collection and telemetry systems for hydrological applications. Spectrascan Limited, 42 New Brighton Road, Emsworth, Hants PO10 7QR (02434 71866).

19. Spembly Ltd

Spembly Ltd has been involved in satellite data reception and processing since 1958, just one year after the launch of the first artificial satellite, SPUTNIK 1. The company currently operates and maintains the tracking station at Lasham, Hampshire, feeding up to 40 outstations on the Meteorological Office network. Spembly personnel are also employed at the National Remote Sensing Centre to assist with data analysis on other types of remote sensing satellites. Meteorological ground stations are built to customers' needs. Spembly Limited, 5 Vicarage Hill, Alton, Hampshire GU34 1HT (0420 88683).

20. Systems Designers Ltd

Systems Designers Ltd has gained considerable experience over many years in real-time on-line computer systems for command, control and communications applications in both defence and industrial markets. The company has provided computer systems for remote telemetry, traffic management, production process control, and telecommunications. In the remote sensing field it has been responsible for the development of specialist software for synthetic aperture radar processing systems.

It is currently contracted to RAE, Farnborough to set up a centre to record and archive data from ERS-1 (European Remote Sensing Satellite), expected to be launched in 1988. Systems Designers Limited (SDL), Systems House, 105, Fleet Road, Fleet, Hants (02514 22161).

21. W Vinten Ltd

The Vinten Group has wide ranging interests covering aerial reconnaissance, light aircraft, exotic materials and coatings, instrumentation systems, electro-optics, computer systems and TV studio equipment.

Vinten has supplied multispectral camera systems to Australia, Canada, ITC (International Institute for Aerial Survey and Earth Sciences) and, via European Community-supported aid progammes, to Third World countries for resource evaluation and planning. W Vinten Limited, Western Way, Bury St Edmunds, Suffolk IP33 3TB (0284 2121).

B. OTHER DETAILS

1. UK Companies Engaged in Supply and Servicing

The following companies are involved in the supply and servicing of scientific instrumentation manufactured overseas.

(i) BIROL: Bristol Industrial and Research Associates Ltd, PO Box 2, Portishead, Bristol BS20 9JB (0272 847787).

(ii) Intertrade Scientific Ltd, Mill House, Boundary Road, Loudwater, High Wycombe, Bucks (06285 28231).

(iii) Flex Systems, John Scott House, Market Street, Bracknell, Berks RG12 1JB (0344 52929).

2. Academic Institutions Offering Consultancy Services

The following academic institutions are equipped to offer consultancy services via expert personnel and advanced analytical and other equipment.

(i) Imperial College of Science and Technology, Centre for Remote Sensing (Blackett Laboratory), Department of Physics, London SW7 2B7 (01-589 5111). Research and assignment work extends to planetary remote sensing, using the IPIPS system, developed by the Planetary Atmospheres Group.

(ii) Nottingham University, Department of Geography, Nottingham NG7 2RD (0602 506101). There is special consultancy support for research in agricultural land use, development planning and coastal environments. Software designed to run on general purpose computers, including micros, has been developed.

(iii) Open University, Department of Earth Sciences, Walton Hall, Milton Keynes, Bucks MK7 6AA (0908 652079). Specialist expertise in geological applications, environmental change detection and photo map production.

(iv) Reading University, Department of Geography, 2 Earley Gate, Reading, Berks RG6 2AU.

(v) Silsoe College: see entry on p. 148. A regional station of the NRSC but with independent teaching, training and research facilities.

(vi) Sheffield University, Image Processing Bureau, Department of Geography, Western Bank, Sheffield S10 2TN. Facilities for image processing for interactive manipulation and access to a range of ancillary reference materials. There is no limitation on application fields.

Other academic institutions also support contract research based on the expertise of staff members. The Remote Sensing Yearbook (1985, 1st edition) provides details of the range of recent and current topics, together with the names of research supervisors.

The interests of companies involved in aerial survey are served by The British Air Survey Association which guarantees the high and consistent standards of its members. The BASA, c/o J A Storey and Partners, 92-94 Church Road, Mitcham, Surrey CR4 3TD (01-640 1971/5).

Part III

Educational and Training Facilities

PART 3.1: EDUCATIONAL ESTABLISHMENTS AND COURSES

In recent years there has been a gradual but progressive expansion of courses specific to remote sensing in UK institutions of tertiary education. The NRSC and NERC also provide, or sponsor, short training courses. Commercial organisations and some public agencies offer courses on various aspects of methodology and data application, some of which may be open to students of remote sensing.

Details of all undergraduate and postgraduate courses, and opportunities for supervised research, are listed in the Department of Trade and Industry's directory (see p. 133). Its intended successor (see p. 133) will bring this fully up to date.

This brief survey of course provision in the UK, valid for the beginning of the 1985/86 academic year, is an attempt to give potential students some guidance on the nature and extent of educational provision. It is not comprehensive, listing only institutions where specialist courses are available. Details can be obtained through direct contact with the institutions in question.

Overseas students should consult the _Report on Full-Time Training Facilities in the United Kingdom in Land Survey, Photogrammetry, Remote Sensing, Cartography and Map Reproduction... which are available to overseas students_, Directorate of Overseas Surveys (Revised Edition, June 1983) 29pp. Copies are held by British Council Offices in most countries.

A. UNIVERSITIES IN THE UK

The department, section or faculty with major responsibility for course provision and development is given in brackets; sometimes courses are available on an interdisciplinary or other co-operative basis.

1. Undergraduate Courses

Aston (Civil Engineering); Aberystwyth UC (Geography); Holloway and New Bedford College, London (Geography); Birkbeck College (University of London); Bristol (Geography); Cambridge (Geography); Dundee (Physics, Electronic Engineering); Durham (Geography); East Anglia (Environmental Science); Edinburgh (Geography); Glasgow (Geography, Topographic Science); Imperial College, London (Physics, Pure and Applied Biology); Keele (Geography); Liverpool (Geography); Newcastle upon Tyne (Geology, Geophysics); Oxford (Atmospheric Physics, Forestry, Geography); Nottingham (Geography); School of Oriental and African Studies, London (Geography); Reading (Geography, Meteorology); Southampton (Geography, Oceanography); Swansea UC (Geography); Sheffield (Geography); Surrey (Electronics and Electronic Engineering); University College, London (Geography, Physics and Astronomy; Photogrammetry and Surveying).

2. Postgraduate Studies

Supervised research leading to MSc, MPhil and PhD degrees is available at all of the above institutions. In most cases, research opportunities reflect the specialist interests of teaching staff.

One-year taught master's degree courses specific to remote sensing are offered by:

(i) University of London. A one-year MSc taught jointly by University and Imperial Colleges, with contributions from other Schools and Colleges within the University. A one-term course is followed by the study of the application of image processing techniques within two specialist options during the second term. The final part of the course involves the presentation of a project report on a chosen topic or theme. The main purpose of the course is to provide trained personnel for commercial and public agencies utilising remote sensing data. It is supported by the NERC and SERC to the extent of providing ten advanced course studentships annually. Further details can be obtained from: Department of Geography, University College London, 26 Bedford Way, London WC1H OAP. Closing date for applications is the end of April for courses starting annually in October.

(ii) University of Nottingham. A one-year MPhil course for graduates with some initial experience of remote sensing. Conventional lectures and seminars are supplemented by research for a major project related to the interests and/or country of origin of candidates. Further details can be obtained from: Department of Geography (Remote Sensing Unit), University of Nottingham, Nottingham NG7 2RD; a booklet on Education and Training in Remote Sensing is available.

(iii) University of Glasgow. Several long-established postgraduate courses viz. (a) Diploma in Cartography; (b) Diploma in Surveying; (c) Diploma in Photogrammetry; (d) MSc in Applied Science (Topographic Science): all courses relate more or less closely to the field of remote sensing, especially (c). Candidates for (d) have the opportunity of presenting a dissertation in a field of specialist interest. A brochure giving full details of courses is available from: Department of Geography, University of Glasgow, Glasgow G12 8QQ. (Glasgow University offers a first degree in Topographic Science, which includes substantial coverage of photogrammetry and remote sensing).

(iv) University of Edinburgh, Department of Geography. MSc and Diploma courses are offered in Geographical Information Systems which include a significant element on remote sensing. (See Addenda.)

(v) University College London. One-year taught MSc and Diploma courses in Photogrammetry and Numerical Methods in Photogrammetry are provided by the Department of Photogrammetry and Surveying, Gower Street, London WC1 6BJ.

(vi) Silsoe College, Cranfield Institute of Technology. An MSc course in Applied Remote Sensing started in October 1985. See p. 148 for further details.

(vii) University of Dundee. A one-year MSc in Remote Sensing, Image Processing and Applications, with interdisciplinary scope and relevance, is offered. The first six months is devoted to formal teaching and practical experience, the second to project work nominated by individual students. There is some bias towards atmospheric and marine applications. A nine-month Diploma in Science course on Digital Mapping and Remote Sensing is also available. Details can be obtained from: Carnegie Laboratory of Physics, University of Dundee, Dundee DD1 4HN.

(viii) University College Swansea. Diploma in Cartography, which includes photogrammetry, in the Department of Geography.

(ix) University of East Anglia. MSc in Resource Assessment for Development Planning, divided into seven units within the framework of 'Land Evaluation'. Remote sensing is an important component in several of the units. Details can be obtained from: UEA, School of Environmental Studies, Norwich NR4 7TJ.

Several additional courses are in active preparation, including a one-year MSc in Remote Sensing at the Department of Earth Science of the Open University. In common with the OU's approach this course will be dependent on distance and interactive learning approaches, utilising printed course materials, TV programmes, radiovision materials, 'face to face' tutorials and practicals. The course will also be open to Associate Students, who register specifically for one course. Full details have yet to be announced. (The Open University can already accept research students working on specific aspects of remote sensing with relevance to earth science).

Other institutions offering opportunities for research in remote sensing applications to graduates include the Scott Polar Research Institute, Lensfield Road, Cambridge and the Macaulay Institute for Soil Research, Craigiebuckler, Aberdeen AB9 2QJ.

B. POLYTECHNICS IN THE UK

Undergraduate courses specialising wholly or in part in remote sensing are currently available at North London (Geography); North-East London (Land Surveying); Portsmouth (Geography); Lancashire (Surveying and Planning); and Kingston (Electronic Engineering and Computer Science).

All of these institutions can accept full or part-time postgraduates undertaking supervised research for higher degrees validated by the Council for National Academic Awards.

Luton College of Higher Education includes a significant element of remote sensing in its BTEC HND in Geographical Techniques (three-year course).

C. SHORT COURSES IN THE UK AND OVERSEAS

1. Silsoe College

The institution with the longest experience in operating short courses in aspects of photogrammetry and remote sensing is Silsoe College, a constituent body of Cranfield Institute of Technology. Silsoe, formerly known as the National College of Agricultural Engineering, offers both short and long courses in a variety of remote sensing contexts. A traditional expertise in air photo analysis and interpretation, especially in the fields of agriculture and soil science, has recently been added to by the acquisition of image processing facilities as part of Silsoe's role as a regional centre of the NRSC. Courses now available range widely within the general context of quantitative analysis of remotely-sensed data applied to natural resources. They are particularly relevant to employees in public and private organisations (both UK and overseas) with responsibility for environmental resource surveys in general. Silsoe can provide courses 'customised' to the specific requirements of small groups, but also organises a programme of standard named courses. It can provide, or draw upon, expertise in geomorphology, geology, soils, forestry, hydrology and water resources, agriculture, and range management. Courses are intensive, practically-oriented and directed strongly towards the application of remote sensing data to natural resource inventory and problem solving. Most are restricted to small numbers of students and vary in length between three days and ten weeks. Many courses are dedicated to remote sensing applications in Third World countries. Full details are available from: The Short Course Secretary, Department of Agricultural Engineering, Silsoe College, Silsoe, Bedford MK45 4DT.

Silsoe College provides one-day sessions (at a frequency of once a month) on Interactive Image Processing using its GEMS facility. A one-day Beginner's Workshop (restricted to a maximum of ten people) is supplemented by a more advanced one-day workshop session, although the two types of courses can be regarded as independent. Additional, more specialised sessions are available for small groups on an ad hoc basis. In all cases, it is possible to utilise imagery of specific interest to participating groups. These workshops are intended as basic familiarisation with 'self-drive' image processing systems and are of potential value to researchers, company employees, staff of teaching and government institutions and consultants. Details can be obtained from: Professional Development Executive, Silsoe College, Silsoe, Bedford MK45 4DT.

2. Other Academic Establishments

The University of Nottingham and North East London Polytechnic (NELP) also offer both standard and flexible short orientation and training courses. NELP courses are specific to the field of land surveying, although the treatment of remote sensing is broadly based; courses at Nottingham are adaptable to discipline interests of participants.

The University of Aston offers Basic and Advanced Training courses in Remote Sensing for small groups of both home and overseas students, varying in duration from one month to three years, according to background and interests. Most students engaged in study for one year or more register for MPhil or PhD degrees. These flexibly organised courses have no formal commencement dates and are therefore adaptable to individual requirements. Further details from Remote Sensing Unit, Department of Civil Engineering, University of Aston, Gosta Green, Birmingham B4 7ET.

The Department of Photogrammetry and Surveying, University College London, arranges courses of one to three days' duration on the principles and applications of photogrammetry for individuals requiring to update their knowledge. Special short courses specific to practical applications are organised on an ad hoc basis, together with opportunities for collaborative research using departmental equipment.

Since 1980, the University of Dundee (Carnegie Laboratory of Physics) has organised and hosted a number of three-four week Postgraduate Summer Schools. They have been sponsored by SERC, NATO, European Space Agency and EARSeL (The European Association of Remote Sensing Laboratories) with material support from the NRSC, and devoted to a specific theme (Meteorology, Oceanography and Hydrology in 1980; Marine Science in 1982; Civil Engineering in 1984). These schools have included lectures, laboratory and computing exercises, image processing experience, audio-visual instruction and field excursions. The school for 1986 will be concerned with Applications in Meteorology and Climatology. The proceedings of each have been published (see p. 68-71). Participants were drawn from a variety of European countries, including the UK, but these schools have also proved attractive to individuals from North America and Third World countries. EARSeL is committed to the provision of further summer schools and other types of courses, through the initiative of its Working Group 3 (Education and Training). EARSeL is a Study Group of the Parliamentary Assembly of the Council of Europe supported by ESA with the objectives of encouraging and co-ordinating research in all aspects of civil remote sensing in western Europe. It is a corporate body offering membership to all institutions and agencies (not individuals) engaged in remote sensing; non-European organisations and commercial organisations are admitted as observers. Members receive a copy of EARSeL News (three issues a year) which provides full details of Working Group activities; carries features on remote sensing research progress; digests recent worldwide developments on all aspects of operational and planned missions, and programmes and profiles the organisational structure of remote sensing in European countries. An annual

Directory is also published (8th Edition, 1985) listing the research activities of contributing members. EARSeL is particularly active at the moment in preparing for publication a variety of teaching materials for use at under and postgraduate levels. Further details from: EARSeL General Secretariat, 292 rue St Martin, F-75003, Paris, France.

Short courses, not part of a continuing programme, in various technical or discipline-specific aspects of remote sensing have been organised in the recent past by: Aston University (land resources and land use planning); University of Southampton (applications in water resources engineering); Lancashire Polytechnic (land surveying); University of Surrey (civil engineering); University of Dundee (coastal and highway engineering; environmental data for offshore development; oceanographic applications; pollution monitoring; land dereliction). The Adult Education or Extra-Mural departments of the Universities of Dundee and Southampton have organised one-day schools, or conferences, for professionals and for general public interest. Further provision of courses may be expected from these and other institutions. Learned societies and professional bodies are also involved in organising one-day or weekend meetings with a focus on remote sensing. The Royal Society has been particularly active in this respect, and publishes the proceedings of relevant meetings in its Philosophical Transactions and in other forms (see Part 2.2 for recent titles).

The National Remote Sensing Centre established a Working Group on Education and Training in 1981. The role of this group, consisting of members from educational, government and commercial establishments, is to advise the Centre on national training needs and to develop ideas and teaching materials. It also attempts to provide teachers, lecturers and others with advice and support in any aspect of their development of teaching and learning strategies in the field of remote sensing. The Secretary of the Group can be contacted via the NRSC (see Addenda).

Remote sensing has yet to make any significant impact on secondary school syllabuses, with the narrow exception of the use of meteorological imagery in A level geography courses. A number of texts, atlas-style workbooks, microprocessor programs and other resources are in active preparation to supplement existing materials. Most of the experience of using remote sensing as a learning medium with younger students has been acquired in Canada and the USA.

The Geography Department of the University of London Institute of Education has an interest in the adoption of satellite imagery in primary and secondary level teaching; a number of theses have been completed and are available on loan from The University Library, 20 Bedford Way, London WC1H OAL (01-636 1500).

3. Commercial Organisations

Certain commercial organisations are also equipped to offer specialised training in one or more of the technologies or methodologies of remote sensing. As with the other courses mentioned earlier, these opportunities are particularly attractive to small groups from developing countries. In all cases, they can be adapted to the background, interests and responsibilities of participating groups. Initial enquiries should be addressed to any of the following:

(i) Masdar (UK) Ltd, Masdar House, 141 Nine Mile Ride, Finchampstead, Wokingham, Berks RG11 4HY (particularly agricultural and land use applications).

(ii) Nigel Press Associates, Old Station Yard, Marlpit Hill, Edenbridge, Kent TN8 5AW.

(iii) Clyde Surveys, Reform Road, Maidenhead, Berks SL6 8BU.

(iv) Hunting Surveys Ltd, Elstree Way, Borehamwood, Herts WD6 1SB.

(v) The Landscape Overview, 26 Cross Street, Moretonhampstead, Devon (residential short courses on air photo interpretation skills).

4. Government Agencies

Government agencies may also sponsor or host training courses for existing staff. Short ad hoc six-eight week training courses for personnel from overseas countries are organised by the Land Resources Development Centre, Overseas Development Administration, Tolworth Tower, Surbiton, Surrey KT6 7DY. Remote sensing is taught as a component of courses on land resource appraisal and land management.

The Natural Environment Research Council (NERC) and the Science and Engineering Research Council (SERC) are the source of most funding for research studentships and training course sponsorship. A recent report (November 1984) has strongly urged that the NERC should have exclusive responsibility for the award of grants in support of remote sensing research in the UK. The NERC maintain advanced digital image processing equipment for use by accredited individuals and groups, and offer short training courses in analytical techniques, as well as other services (see pp. 34-35 for further details).

5. Organisations in Western Europe

In Western Europe, educational and training opportunities are numerous. Some courses are given largely, or entirely, in English when they attract international participation. The availability of many courses is advertised in the Remote Sensing Society's News and Letters and EARSeL News. Apart from EARSeL-initiated conferences and Summer Schools (which may grow in frequency and ambition), it is also worth noting:

1. Joint Research Centre (JRC) of the Commission of the European Community (Ispra, Italy). JRC organise an annual programme of scientific and technological training courses, in which remote sensing has featured strongly in recent

LIVERPOOL INSTITUTE OF
HIGHER EDUCATION
THE BECK LIBRARY

years. Courses are between one and four weeks in duration, and are normally given in English. They are open to teachers, postgraduates, research workers, scientific and technical staff in public agencies and institutions in EEC and developing countries. Most courses presuppose some relevant scientific training, and place strong emphasis on practical experience. Numbers are normally limited to 15-20 participants and are held in September or October. Details (including the availability of financial support) are obtainable from: ISPRA, Courses Secretariat, Centro Comune di Ricerca, 21020 Ispra (Varese), Italy. Recent courses have included: Remote Sensing for Land Use Inventories (1982); Synthetic Aperture Radar, Principles and Applications (1983) and Remote Sensing Image Processing Applications (1984). JRC co-operates closely with the Council of Europe (EARSeL); ITC and other European organisations in this field.

2. NATO Advanced Studies Institutes Over 100 so-called institutes are organised annually, most of them very specific. The object of each is to publicise new developments and foster contacts between practising scientists and research students, drawn from NATO countries. Most are courses of two weeks' duration involving up to 100 participants. Sponsorship, and modest financial assistance is available from: NATO Scientific Affairs Division, B-1110, Brussels, Belgium. In recent years, remote sensing topics have been included in all programmes, and have attracted contributors conversant with the 'state of the art' of their specialisations. The proceedings of most institutes are subsequently published by D Reidel and Plenum Press. Details of forthcoming meetings can be obtained from the above address.

3. GDTA (Groupement pour le Développement de la Télédétection Aérospatiale) GDTA offer a semi-continuous programme of eight-ten week training courses in the full range of concepts and methods of remote sensing, with the collective title of Remote Sensing Training Sessions, and under the aegis of Ecole Nationale des Sciences Géographiques (Paris). Most courses are appropriate to professionals, including teachers and lecturers, without any prior familiarity with remote sensing, and are organised on a modular basis. Courses are repeated, according to demand, but not all are in English. Courses can be organised on behalf of specific training groups and independent organisations. The vigour of development of remote sensing technology in France in recent years gives these courses a particular flavour. GDTA is a framework organisation that co-ordinates expertise in teaching and training from leading French institutions, notably the Institut Géographique Nationale; Centre National d'Etudes Spatiales; Bureau pour le Développement de la Production Agricole, and the Bureau de Récherches Géologiques et Minières. Most courses are located at Toulouse and provide opportunities for visits to the important CNES facilities there. Full details of courses can be obtained from: GDTA, 18 avenue Edouard Belin, 31055, Toulouse, France. Costs are normally borne by participants or their employers.

4. ITC (International Institute for Aerospace Survey and Earth Sciences), The Netherlands, has a long and highly respected tradition of provision of both postgraduate and professional training in remote sensing. Established in 1950 its courses and research interests are directed towards developing countries but opportunities for students and professional employees from elsewhere can be individually negotiated. ITC also functions as a consulting agency and provides direct support to national remote sensing organisations in India, Colombia, Indonesia and Nigeria. Its staff have experience in the application of remote sensing methods in a very wide range of environments and environmental problems. Many courses require a BSc degree for admission and lead to the award of master's degrees, post-graduate diploma and technical certificates, and vary in length between 12 and 18 months; some are designed for basic training only and are not linked to specific awards. Participants may qualify for financial assistance through The Netherlands' government Programme for International Development. Individual students have the potential opportunity to pursue any one of over 50 structured courses at their own pace, under individual supervision. The whole programme is comprehensive but flexible. Any one of several alternative courses can be taken in Photogrammetry; Aerial Photography (in General); Cartography; Remote Sensing and Geology; Applied Geology; Remote Sensing and Geomorphology; Remote Sensing and Soil Survey; and Remote Sensing and Forestry; Integrated Natural Resources Surveys; Rural Land Use and Rural Resources; Urban Surveys; Geographical Landscape Analysis; Applied Geomorphology; Multi-Disciplinary Surveys for Development Planning; Mineral Exploration and Exploration Geophysics, and Land Information Systems,all with reference to aerial survey and remote sensing techniques. Each course is described in detail in individual brochures available from ITC. They are given in English. Not all of them are offered simultaneously, of course. Full information is available from: Student Affairs Office, ITC, 350 Boulevard 1945, PO Box 6, 7500 AA Enschede, The Netherlands (053 320330).

It should be mentioned here that ITC contributes to seminars and workshops organised by the United Nations, and has specific experience in designing and conducting training courses in developing countries, in co-operation with both national and international agencies. Short courses designed to update and extend the knowledge of professionals already working in natural resources survey fields are organised on a regular basis. The approach in all cases is essentially practical.

6. Organisations in North America

Large numbers of professional and other training courses are offered by numerous North American institutions, both academic and governmental. No attempt can be made here to provide a comprehensive list, but students with an interest in undertaking postgraduate research in the USA and Canada should consult the paper by L D Nealy in Photogrammetric Engineering and Remote Sensing 1977, 43(3), 259-284. More up to date listings may be available from the American Society of Photogrammetry and the Association of American Geographers. Forthcoming training courses in the USA are listed in the Landsat Data Users Notes (see p. 54).

One-month courses and workshops are organised by the US Geological Survey, in co-operation with other institutions, for non-US scientists, resource managers and engineers. Some are of a basic orientation nature, others are discipline-specific or devoted to analytical techniques. Despite the title of the organising agency, the majority of courses are not confined to (or even relate to) geology. There have been two training courses specifically for foreign participants, at the EROS Data Center (Sioux Falls, South Dakota) every year since 1973. Students are not required to have any technical or scientific background in the subject. More advanced courses are organised by the USGS in co-operation with Northern Arizona University (Flagstaff, Arizona).

Advanced courses offered in recent years have been concerned with geological interpretation, land-use planning, digital processing for earth scientists, geological hazards, photographic technology, hydrological exploration and water resources planning, and vegetation assessment. For some courses prerequisite experience and/or qualifications may be defined. Full details of these courses and workshops are available from: Chief, Office of International Geology, US Geological Survey, 917, National Center, Reston, Virginia 22092, USA.

The Laboratory for Applications of Remote Sensing (LARS), Purdue University has carried out a pioneering and innovating role in the provision of a wide variety of courses of an introductory and more advanced nature. There is a well-established short course on Remote Sensing Technology and Applications, which runs for four days and is offered monthly. It is designed to familiarise participants with the physical basis of the subject and its potential application to their fields of interest. LARS is also a major centre for the innovation of teaching materials, especially self-instruction media. Its staff conduct research and consultancy and publish extensively. Details of courses, publications and facilities can be obtained from: LARS, Purdue University, 1220 Potter Drive, West Lafayette, Indiana 47906, USA.

LARS also administers a Visiting Scientist Program, providing an opportunity for individual scientists to become acquainted with any area in which LARS has developed expertise, such as crop inventory systems, ecosystems research, data processing and technology transfer. Costs are paid by trainees or their sponsoring agencies. Details can be obtained from the above address.

This is only the 'tip of the iceberg' so far as North American provision is concerned. Unfortunately, there is no published directory of institutions for the guidance of students, although a consolidated list of academic and professional courses in Canada is available from the Canada Centre for Remote Sensing. Particularly noteworthy are the numerous short training courses in data processing and resource surveys provided by the provincial centres, e.g. the Ontario Remote Sensing Centre at Toronto and the Alberta Centre at Calgary.

7. Other Organisations

The Soviet Union and East European States engage in a wide variety of research, development and application project work, some of it through INTERCOSMOS (Council on International Co-operation in the Study and Utilisation of Outer Space), with its headquarters in Moscow. National and international collaboration on education and training in remote sensing is pursued, with some opportunities for involvement by non-socialist states. Nearly all courses are discipline-related and give particular emphasis to the integration of remote sensing and ground-based survey and analysis. Details of opportunities are not usually circulated outside the INTERCOSMOS 'catchment', but further information may be obtained from: INTERCOSMOS, 117901 Moscow, V-71 Leninsky Prospect 14, Moscow, USSR.

The United Nations sponsors or supports a varied programme of training courses through its constituent organisations. These are normally advertised in advance in *Nature and Resources*, published quarterly by UNESCO (Paris). There are two particularly active centres for the provision of training in remote sensing methodology, both dedicated to the needs of developing countries, viz:

(i) Remote Sensing Unit, Natural Resources and Energy Division (DNRE) New York, USA.

(ii) UN Remote Sensing Centre, Food and Agriculture Organisation, Via delle Terme di Caracalla, 00100 Roma, Italy.

Both provide technical advice and project assistance, as well as problem-oriented and discipline-specific training. The FAO also convenes international and regional seminars and workshops for expert advisors and others concerned with organising training courses. An important objective of both centres is to assist in the development of regional and national remote sensing activities in developing states. In recent years, FAO activities have emphasised desertification and agro-ecological monitoring, e.g. pasture and range land management. Interests in utilising remote sensing in natural hazard prediction have been a feature of recent institutional development, as well as assisting with forest and fisheries resource inventories.

Both centres co-operate with the UN Outer Space Affairs Division (UNOSAD) (New York). The FAO liaises closely with ESA, WMO, ITC and UNDRO (UN Disaster Relief Organisation); DNRE with the UN Environment Program (UNEP), UNESCO, the UN Development Program (UNDP) and the USGS. At the time of writing, a major global programme to expand remote sensing training is under consideration through the auspices of the UNDP. The UN Economic and Social Affairs Division operates several Regional Commissions (Asia and the Pacific; Africa; Latin America and the Caribbean; the Middle East), each of which operates a remote sensing program. These have been concerned with promoting conferences and meetings to familiarise technical personnel with developments in the subject. It is anticipated that the UNDP initiative referred to above will absorb this activity. The case for improved integration and rationalisation of UNO involvement in remote sensing is evident; UNOSAD and the UN Committee on the Peaceful Uses of Outer Space are not managerial bureaucracies, but rather standing committees concerned with the resolution of international legal problems resulting from both remote sensing and telecommunications satellites and spacecraft. Periodically, the UN organises 'jamboree' style conferences on behalf of all member states (dubbed UNISPACE) at which the many facets of intergovernmental activities in space science are represented. Remote sensing has formed a significant element of these proceedings, with the generation of a number of 'position' and 'review' papers. In the field of education and training, the staff of ITC have prepared discussion papers on the procedures of technology and training transfer from developed to developing states.

Careers in remote sensing

Most of the opportunities for full-time careers in remote sensing technology and/or its applications are provided by the public agencies, educational establishments and commercial companies named in this text. To these might also be added: British Petroleum (Research); British Aerospace; Plessey; IBM (UK); Marconi Space Systems, Portsmouth; Marconi Radar Systems, Leicester; Geomorphological Services (GSL); Marlow and the Rutherford Appleton Laboratories (SERC), Didcot. The Ministry of Defence is the single largest employer of trained remote sensing specialists, in units such as the Space Department, RAE Farnborough; Admiralty Research Establishment (Portsmouth; Portland); Mapping and Charting Establishment (Feltham, Middlesex) and JARIC (RAF Brampton, Huntingdon). Personnel are recruited through the Civil Service Commission. In addition, the armed services also directly recruit personnel for both active duty and training functions in fields that relate directly or partially to remote sensing.

Appointments in the computing industry, information technology, general space technology and management/administration of research may be suitable to graduates with a training in remote sensing.

PART 3.2: LEARNED SOCIETIES IN THE UK

A. The Remote Sensing Society

For anyone in the UK seeking to enhance their interest in remote sensing, membership of the Remote Sensing Society is highly recommended.

Founded in 1974, the RSS is the only learned society in the UK devoted to the advancement of all aspects of the subject. Although based in the UK, it has always pursued an international outlook; recently branches of the Society have been established in certain other countries for the purpose of holding meetings, organising lectures, etc.

The main activities are:

- Organising a calendar of lectures, technical meetings, workshops, etc on a variety of themes and topics consistent with the wide spectrum of interests of members. (Some meetings are arranged in co-operation with other societies and organisations.)

- Convening an annual conference, normally dedicated to a broadly defined theme, in August or September of each year.

- Publishing proceedings of its meetings, symposia, conferences, etc.

- Editing the International Journal of Remote Sensing, and producing the Remote Sensing Society's News and Letters (the latter is quarterly, devoted to short contributions and characterised by a rapid interval between manuscript submission and publication); it carries news items, reprinted from the IJRS. (Both are published by Taylor and Francis, Basingstoke, Hants.)

- Making an annual award for scientific excellence of published material.

Meetings are held in a variety of locations, with a bias towards London that reflects the distribution of members and the accessibility of the capital. The annual conference is peripatetic.

In addition, the Society is committed to the encouragement of educational, commercial and governmental use of, and participation in, remote sensing activities, particularly in the UK. It acts as an advisory source for individuals and corporate organisations seeking both general and specific assistance in any context relating to remote sensing.

The Society maintains a library available to members for the consultation and loan of items. It also publishes an informative News and Letters (four issues a year), which reviews new developments, lists new publications and forthcoming meetings (worldwide) and reports news and views from members.

Membership is available in four categories, viz:

(i) Ordinary, for persons with a professional interest or active engagement in remote sensing. Benefits include free receipt of News and Letters, annual conference proceedings, and attendance at Society meetings at no, or reduced, cost. Other publications and subscription to the International Journal of Remote Sensing, are available at favourable rates (cost: £18.00).

(ii) Student, open to students undertaking recognised undergraduate, postgraduate and other courses. Same privileges as Ordinary members, with the exception of receipt of the annual conference proceedings (cost: £6.00).

(iii) Corporate, available to government, publicly funded and commercial organisations with an interest in remote sensing. Corporate members do not have voting rights at General Meetings, but enjoy all other membership benefits (cost: £84.00).

(iv) Affiliation, for educational institutions, libraries, learned societies and all non-profit making organisations. Publications are received on the same basis as Ordinary members, but there are restrictions on the number of persons able to attend meetings (cost: £36.00).

At the time of writing, the Society is considering the institution of professional grades of membership, including a Fellowship category.

The Remote Sensing Society is administered by a Council of 12 elected members, assisted by co-opted members and an administrative secretary. The current President is Sir Hermann Bondi. In common with most learned societies, its development is dependent on the voluntary, informal input of many individuals. The Society has the unusual distinction of counting most people active in the subject amongst its membership; its conference and meetings therefore encourage the rapid exchange of experience, advice, etc. within the UK. A further strength of the Society is the balance of membership drawn from the government, industry and educational sectors. It is therefore in a strong position to identify the arguments, attitudes and activities of the remote sensing community in the UK, and (to a lesser extent) overseas. This role has been recently demonstrated by the Society's representations to Parliamentary committees and its active involvement in attempts to create more comprehensive forms of international co-operation. To this end, it has established a panel of Regional Meetings' Secretaries in over 20 countries.

Full details of the Society are available from: The Secretary, Remote Sensing Society, c/o Department of Geography, 2 Earley Gate, University of Reading, Reading RG6 2AU (0734 665633). An attractive brochure describing the Society, and a free copy of a recent edition of News and Letters are available on request.

B. Other Societies

The British Interplanetary Society, established in 1933, is a society that serves professional interests and also promotes the popularity of astronomy and space science for the general public, school children and students. Its activities are diverse and include remote sensing in the broadest sense. To this end, it organises technical symposia, seminars and conferences at various levels and publishes relevant papers in the monthly Journal (JBIS), which is devoted to technical articles with individual issues often confined to specific themes; Spaceflight, with a magazine-style format, appears ten times per annum, and contains a digest of news, views, book reviews and features on current and future space projects; and Space Education (two issues per annum) is dedicated to articles, review and news sections of relevance and interest to secondary school pupils and teachers. The latter also have the opportunity of hiring films and videos.

In addition, the Society markets special books, slide sets and items such as badges, ties, sweatshirts and commemorative souvenirs. Its outlook is international (over one third of members are from outside the UK) and it has the distinction of being the oldest established society devoted exclusively to space science. Membership is according to age status (in 1985, £16.00 per annum for under 18s; £18.00 per annum 18-20; £21.00 per annum 21-65, and £19.00 per annum over 65). Members receive Spaceflight and an opportunity to subscribe to the Society's other journals at reduced rates; free attendance at lectures, meetings and film shows; opportunities to participate in specially organised visits and access to the Society's extensive library. Fellowship is by election. Full details available from the Executive Secretary, British Interplanetary Society, 27-29 South Lambeth Road, London SW8 1SZ.

The Photogrammetric Society publishes The Photogrammetric Record and organises an annual programme of lectures, workshops and conferences (often in co-operation with other learned societies). Ordinary membership is £12.50 and Junior membership £7.00 per annum. Details can be obtained from Department of Photogrammetry and Surveying, University College London, Gower Street, London, WC1E 6BT. The Society has established a Working Group for Photo-Interpretation and Remote Sensing, with terms of reference extending to education and training.

Numerous learned societies and other professional organisations arrange meetings (of a varied nature) that may be wholly, partly or peripherally related to remote sensing. In recent years these have included the Royal Society, Geological Society, Institute of British Geographers, Geographical Association, Royal Meteorological Society, British Pattern Recognition Society, Royal Institution of Chartered Surveyors, Royal Geographical Society, British Geomorphological Research Group and several others. Their meetings are usually publicised in advance in calendars of forthcoming conferences, etc appearing in the News and Letters of the Remote Sensing Society, together with contact addresses for further details. Certain meetings may be restricted on the basis of numbers or membership. Government agencies, notably the NRSC and NERC, also organise and host meetings, usually on a specific topic or theme on an irregular basis. These meetings are often co-operative between different groups.

Part IV
Further Information

PART 4.1: LIST OF ABBREVIATIONS AND ACRONYMS USED IN REMOTE SENSING

ASP	American Society of Photogrammetry.
AVHRR	Advanced Very High Resolution Radiometer.
BGS	British Geological Survey.
BIS	British Interplanetary Society.
BNCSR	British National Committee for Space Research.
BUFVC	British Universities Film and Video Council.
BL	British Library.
B/W	Black and White (photographic prints).
COSPAR	(International) Committee for Space Research.
CCRS	Canadian Centre for Remote Sensing.
CERMA	Center of Earth Resources Management Applications (USA).
CCT	Computer Compatible Tape.
CNES	Centre National d'Etudes Spatiales (France).
COES	(International) Committee for Earth Observation Satellites.
CZCS	Coastal Zone Colour Scanner (Nimbus 7).
D of E/DOE	Department of the Environment.
DFVLR	German Aerospace Research Establishment (Oberpfaffenhofen, West Germany).
DTI	Department of Trade and Industry.
EARSeL	European Association of Remote Sensing Laboratories.
EDC	EROS Data Center (USA).
EDIS	Environmental Data and Information Service (NOAA) (USA).
ERIM	Environmental Research Institute, University of Michigan (USA).
EROS	Earth Resources Observation System (USA).
ESA	European Space Agency.

ESOC	European Space Operations Centre (Darmstadt, West Germany).
ESRO	European Space Research Organization.
FAO	Food and Agriculture Organisation (UN).
GDTA	Groupement pour le Developpement de la Teledetection Aerospatiale (France).
GOES	Geostationary Operational Environmental Satellite.
HCMM	Heat Capacity Mapping Mission (Satellite).
HMSO	Her Majesty's Stationary Office (London).
IAA	International Aerospace Abstracts (USA).
ICSU	International Council of Scientific Unions.
IEEE	Institute of Electrical and Electronic Engineers (USA).
IGN	Institut Géographique National (Paris; Brussels).
IGARSS	International Geoscience and Remote Sensing Society (USA).
IPIPS	Interactive Planetary Image Processing System (University of London).
IR	Infra Red.
IOS	Institute of Oceanographic Sciences.
IRS	Information Retrieval Service (European Space Agency).
ISPRS	International Society for Photogrammetry and Remote Sensing.
ITC	International Training Centre (Institute for Aerospace Survey and Earth Sciences, The Netherlands).
ITE	Institute of Terrestrial Ecology.
INTERCOSMOS	Council of International Co-operation in the Study and Utilization of Outer Space (USSR).
ITCZ	Intertropical Convergence Zone.
JPL	Jet Propulsion Laboratory (Pasadena, California, USA).
JRC	Joint Research Centre (Ispra, Italy) (European Community).

LACIE	Large Area Crop Inventory Experiment.
LARS	Laboratory for Application of Remote Sensing (Purdue University, USA).
LEDA	On-Line Earthnet Data Availability.
LFC	Large Format Camera (NASA).
MC	Metric Camera (ESA).
MOMS	Modular Optoelectronic Multispectral Scanner.
MSS	Multi-spectral Scanner or Multi-spectral Scanning.
NAPL	National Air Photo Library (Canada).
NASA	National Aeronautics and Space Administration (USA).
NERC	Natural Environment Research Council.
NESDIS	National Environmental Satellite Data and Information Services (USA).
NOAA	National Oceanic and Atmospheric Administration (USA).
NPA	Nigel Press Associates, Ltd.
NPOC	National Point of Contact (Earthnet).
NRSC	National Remote Sensing Centre (UK).
NSTL	National Space Technology Laboratories (USA).
NTIS	National Technical Information Service (USA).
OSTA	Office of Space and Terrestrial Applications (USA).
OU	Open University.
RADAR	Radio Detection and Ranging.
RAE	Royal Aircraft Establishment (Farnborough, Hampshire).
RBV	Return Beam Vidicon (Landsats 1 - 3).
RESORS	Remote Sensing On-line Retrieval System.
RSL	Remote Sensing Laboratory, University of Kansas (USA).
RSS	Remote Sensing Society.
SAR	Synthetic Aperture Radar.

SERC	Science and Engineering Research Council.
SIR	Shuttle Imaging Radar.
SLAR	Sideways-Looking Airborne Radar.
SMS	Synchronous Meteorological Satellite.
SPIE	Society of Photo-Optical Instrumentation Engineers (USA).
SPOT	Satellite Probatoire de l'Observation de la Terre (France).
STAR	Scientific and Technical Aerospace Reports (NASA) (USA).
TAC	Technology Applications Center (USA).
TIROS	Television Infra-Red Observation Satellite.
TM	Thematic Mapper (Landsats 4 and 5).
UN	United Nations Organisation.
UNDP	United Nations Development Programme.
UNDRO	United Nations Disaster Relief Organisation.
UNESCO	United Nations Educational, Scientific and Cultural Organisation.
UOSAT	University of Surrey Satellite Project.
USGPO	United States Government Printing Office.
USGS	United States Geological Survey.
UTM	Universal Transverse Mercator projection.
VHRR	Very High Resolution Radiometer.
WMO	World Meteorological Organisation.

ADDENDA

p43: Because of recent (1985) increases in the costs of obtaining Landsat Thematic Mapper data, the NRSC may not be able to continue its policy of archiving all cloud-free imagery of the British Isles. The same financial constraint will almost certainly apply to the acquisition of SPOT data.

p52: Transfer of the Landsat programme to commercial ownership. The U.S. government finalised the arrangements for private sector operation of Landsat in September 1985. Although data will continue to be available through the EROS Data Center, ownership now passes to EOSAT (The Earth Observation Satellite Company). New product order forms, pricing schedules and details of copyright regulations have been prepared and circulated to all existing EROS customers. As was widely anticipated, the terms of the company's 'Agreement for Purchase and Protection of Satellite Data' introduces restrictions on the use, dissemination and reproduction of Landsat products that did not apply when they were public property. The enabling legislation is the Land Remote Sensing Act (US Public Law 98-365), 1984. This act also guarantees non-discriminatory access to Landsat data for all requesters. Thus, the spirit, if not the detailed practice, of the former open skies policy of the US federal government, is preserved. Details may be obtained from: EOSAT, 8201 Corporate Drive, Suite 450, MetroPlex 11, Landover, Maryland 20785, USA. (Telephone number for non-US enquiries is (605) 594-2291.) Within three years, EOSAT plan to replace the EROS facility with a re-designed and relocated centralised data recording, processing and sales centre. The company also acquires responsibility for all aspects of the future planning of the Landsat system.

p57: The final proof check of this text was completed within days of the anticipated launch of SPOT-1. Full details of the organisational structure for handling SPOT data and of the products and services that will be available have been documented by SPOT-IMAGE. They are outlined in SPOT Newsletter No. 7, July 1985. The full list of national distributors is also given which, for the UK, includes Nigel Press Associates as well as the NRSC. Pro forma for search, quotation and ordering are available, together with prices and details of procedures.

p68: Although of restricted regional scope, a publication that effectively discusses and illustrates the combination of aerial photography and Landsat imagery in the context of resource mapping in a developing country is R. B. King (1984), Remote Sensing Manual of Tanzania, 206pp, $20.00. Published by the Land Resources Development Centre, Overseas Development Administration (Surbiton, Surrey). It provides a model for the evaluation of remote sensing capabilities and needs in Third World states in general.

p82: Remote Sensing On-line Retrieval System (RESORS) is a computerised bibliographic information service operated on behalf of the Canada Centre for Remote Sensing by Gregory Geoscience. Publications are entered into a 'master' index through the use of over 1700 selected keywords. The system is restricted to worldwide terrestrial, atmospheric and oceanic applications of remote sensing and to image processing. RESORS can be directly accessed via the DATAPAC network in Canada, or by making a request for information in writing, by telex or telephone. Further details from: RESORS, Canada Centre for Remote Sensing, (5th Floor) 240 Bank Street, Ottawa, Ontario, Canada KIA OY7 (Telex: 053377; Telephone: 613-995-5645).

p.146: From October 1985, the University of Edinburgh is offering MSc and Diploma courses in Remote Sensing and Image Processing. This is a modular course, with contributions from various departments, taking the form of six months of formal teaching followed by six months of project work. Applicants would normally be expected to have a first degree in engineering, physics or mathematics. A variety of options in the broad field of information technology is available for study alongside the core content of the course. Further details from: Head of Department of Meteorology, University of Edinburgh, King's Buildings, Mayfield Road, Edinburgh.

p150: A course without any equivalent in the UK (though available in North America) is Remote Sensing and Geopolitics. It is part of the Master of Arts programme in Diplomatic Studies organised by the International Studies Unit of the University of Salford and taught at Bedford College, London. (Bedford College is no longer a constituent part of the University of London; it is now known as Holloway and New Bedford College, and is located at Egham.) Details of this course from: International Studies Unit, University of Salford at Bedford College, Regent's Park, London NW1 4NS (Telephone: 01-486 5694).

Index of Authors and Organisations

Underlined entries provide details of the addresses of organisations.